CEREBRAL GLIOMAS

CEREBRAL GLIOMAS

Proceedings of the International Workshop on Brain Tumors, held in Santa Margherita, Ligure, Italy, 20–22 June 1988

Editors:

GIOVANNI BROGGI
Istituto Neurologico 'C. Besta'
Milano, Italy

MASSIMO A. GEROSA
Dipartimento di Neurochirurgia
Verona, Italy

1989

EXCERPTA MEDICA, Amsterdam - New York - Oxford

No responsibility is assumed by the Publisher for any injury and/or damage to persons or property as a matter of products liability, negligence or otherwise, or from any use or operation of any methods, products, instructions or ideas contained in the material herein. Because of rapid advances in the medical sciences, the Publisher recommends that independent verification of diagnoses and drug dosages should be made.

International Congress Series No. 828
ISBN 0 444 81081 1

This book is printed on acid-free paper.

Published by:
Elsevier Science Publishers B.V.
(Biomedical Division)
P.O. Box 211
1000 AE Amsterdam
The Netherlands

Sole distributors for the USA and Canada:
Elsevier Science Publishing Company Inc.
655 Avenue of the Americas
New York, NY 10010
USA

Library of Congress Cataloging in Publication Data:

International Workshop on Brain Tumors (1988 : Santa Margherita
Ligure, Italy)
Cerebral gliomas : proceedings of the International Workshop on
Brain Tumors, held in Santa Margherita Ligure, Italy, 20-22 June
1988 / editors: Giovanni Broggi, Massimo A. Gerosa.
p. cm. -- (International congress series ; no. 828)
Includes bibliographies and indexes.
ISBN 0-444-81081-1 (U.S.)
1. Gliomas--Congresses. 2. Cerebral cortex--Tumors--Congresses.
I. Broggi, G. (Giovanni) II. Gerosa, M. A. III. Title.
IV. Series.
[DNLM: 1. Brain Neoplasms--congresses. 2. Glioma--congresses.
W3 EX89 no. 828 / WL 358 I607 1988c]
RC280.B7I57 1988
616.99'281--dc19
DNLM/DLC
for Library of Congress 89-1379
CIP

Printed in The Netherlands

PREFACE

The aim of this volume is strictly related to the goal of the 'Continuing Education Courses' of Santa Margherita, upon which these Proceedings are based: to represent an update on the subject, as didactical as possible, and to provide an opportunity for scientific interrelation between basic scientists and clinically-oriented specialists. Such a joint cooperative effort should play a major role in devising newer investigative approaches and innovative treatments for cerebral gliomas.

Several of the most relevant and intriguing topics in neuro-oncology have been addressed. As regards the biopathology of these tumors, a special focus has been dedicated to oncogenes and viral neuro-oncogenesis, to biopredictive markers and neurobiological advances. As regards the most recent diagnostic possibilities, PET-scan detection of tumor cell infiltrates and MRI-guided spectroscopy of intracranial gliomas are discussed.

Finally, some meaningful experiences with computerized stereotactic surgery, brachytherapy, radiosurgery, intrarterial and intratumoral chemotherapy are presented.

Tumor grading, the histological, radiological and surgical definition of tumor boundaries and peritumoral infiltrates still represent a major challenge for stereotactic biopsies and for the proper planning of neurosurgical treatments in these patients. However, stereotactic surgery, properly used immunohistochemical–immunobiological techniques, together with computer-assisted tridimensional imaging of cerebral gliomas, are currently providing deeper insights into the biological behavior of these tumors, as well as more appropriate and effective weapons for their therapy.

Our hope and belief, and our strongest commitment to brain tumor patients is to explore all the reliable tools, not only to enable us to treat malignant gliomas, but to find a cure in the nearest possible future. Quite a demanding task, but also a fascinating human and scientific experience.

GIOVANNI BROGGI and MASSIMO GEROSA

CONTENTS

BIOPATHOLOGY

EPIDEMIOLOGY

DIAGNOSIS

TREATMENT

BIOPATHOLOGY

Cerebral gliomas. G. Broggi and M.A. Gerosa, editors

THE GRADING AND TOPOGRAPHY OF ASTROCYTOMAS - AN UPDATE

PETER C. BURGER
Box 2712, Department of Pathology, Duke University Medical Center, Durham, North Carolina 27710

INTRODUCTION

For the immediate purpose of treating an individual patient with a glioma, or for the broader intent of comparing the results of new treatments between large groups of patients in cooperative clinical trials, it is essential that the histologic grade or grades of the neoplasm(s) be well characterized. It is apparent that a variety of grading systems are presently utilized for both of these purposes and it is therefore difficult to compare results of therapies or to establish which therapy is most appropriate for a given patient. The first part of this report reviews the current grading systems for astrocytic neoplasms and focuses on newer methods that may provide more objective measures to predict a tumor's biologic behavior and response to therapy.

In light of the increasing interest in interstitial radiotherapy and intraarterial chemotherapy, the understanding of the topographic anatomy of a gliomas is becoming increasingly important. Although computerized tomographic (CT) images are widely used in dosimetry and in determination of the site of arterial injection, recent evidence suggests that the topographic anatomy of malignant gliomas is considerably more complicated than is apparent by computerized tomography. The second part of this paper reviews the topographic anatomy of malignant gliomas in light of their anatomy as revealed by CT and magnetic resonance imaging (MRI).

GRADING OF ASTROCYTOMAS

The Kernohan system has had considerable influence on the grading of astrocytic gliomas and is still in wide use, particularly in North America. This system was applied to a series of patients with astrocytic neoplasms treated at the Mayo Clinic and first published in 1949 (1,2). Based on the four-tiered Broder's system as originally applied to squamous cell carcinomas of the lip, it assigned astrocytomas into grades 1, 2, 3, and 4. Faithful to its parent system, it categorized these lesions in part on the relative proportions of well differentiated and poorly differentiated cells. This is in contrast to the general present practice of defining a neoplasm on the basis of its most malignant area. In addition, the extent to which the data from the original Kernohan system can be compared to patients of the present era remains questionable since the grade 1 neoplasms included some cerebellar astrocytomas. These are morphologically and biologically distinct from the fibrillary or diffuse astrocytomas in question. In addition, the treatment of gliomas has undergone considerable modification since the time of the Kernohan. Radiotherapy is now used frequently and in higher doses than it was then. In addition, therapeutic means to prolong life such as use of steroids are now available and were not for the patients in the Kernohan studies. Nevertheless, when applied to the material from the Mayo Clinic, the Kernohan system established a close relationship between grade and survival. As is discussed below, there were also close relationships between the grade of the lesion, the age of the patient, and the duration of preoperative symptoms.

The utility of the Kernohan system as originally published has been challenged by many pathologists who have failed to find utility in subdividing the glioblastomas into grades 3 and grade 4. This is well illustrated in a report from the Radiation Therapy Oncology Group (RTOG) that found no statistically significant difference in survival between patients with astrocytomas grade 3 and grade 4 as classified by referring pathologists (3). Thus, the Kernohan system has had a major impact on the grading of gliomas and is still used in modified form. It does not, however, have the influence that it did in the years immediately following its initial publication.

Shortly after the publication of the Kernohan classification, Ringertz suggested a simplified system of three grades: astrocytoma, an intermediate lesion, and the glioblastoma multiforme (4). The survival curves illustrated from these types showed significant differences. The World Health Organization (WHO) classification of 1979 also employed a three-tiered classification of the fibrillary astrocytic neoplasms although the glioblastoma was not considered an astrocytic tumor but was combined with the medulloblastoma in a group of embryonal and undifferentiated neoplasms (5). By the WHO classification, the well differentiated astrocytoma was a grade II, the glioblastoma was a grade IV. An intermediate lesion, the malignant astrocytoma, was a grade III. Grade I was assigned to the pilocytic neoplasms of the cerebellum, hypothalamus, and optic nerve.

There was no data presented at the time of publication of the WHO volume to support the utility of the three-tiered system, although subsequent studies from the EORTC utilizing a modification of the system showed differences in survival between tumors of the various grades (6). Specifically, as reported by Brücher, differing survival curves noted between patients with neoplasms grades III and with IV. It is of special interest that the distinction between these two grades was made partially on topographic grounds in accord with the statement in the WHO classification that the anaplastic astrocytoma is "an astrocytoma of one of the recognized subtypes containing areas of anaplasia." In this use of the WHO system, the anaplastic astrocytoma is conceived as a neoplasm with foci of anaplasia in the background of better differentiated tumor; whereas the glioblastoma is entirely anaplastic. It would appear that many of the grade III lesions in the EORTC study would be considered glioblastomas in other systems where neoplasms are graded on the basis of the most malignant area. It is appropriate that the survival curves of these patients are somewhat lower than grade III neoplasms in other classifications.

A grading system essentially shared by two large cooperative groups in the United States has been reported separately (3,7). This system utilizes the basic three-tiered Ringertz system and gives special weight to the presence of necrosis in distinguishing the glioblastoma multiforme from the anaplastic astrocytoma. The survival curves for these patients with glioblastomas are essentially identical although the survival for the patients with the grade III neoplasm in the RTOG are somewhat better than those in the Brain Tumor Study Group (BTSG). This latter discordance emphasizes the difficulty of comparing survivals of patients between two cooperative groups, even where the classification systems are ostensibly the same. It also underscores the particular difficulty in the subjective distinction between a well differentiated astrocytoma (grade II) and an anaplastic astrocytoma (grade III) where no more objective feature such as necrosis is used.

A system applied in patients at the Brain Tumor Research Center at the University of California in San Francisco divides astrocytic neoplasms into moderately differentiated astrocytoma, moderately anaplastic astrocytoma, highly anaplastic astrocytoma, gemistocytic astrocytoma, and glioblastoma multiforme. The morphologic details of this classification have not yet been published and it is difficult therefore to compare the survival of patients in this institution with those of other cooperative groups.

A recent classification of considerable interest returns to a four-tiered system, although one that is formulated in a means considerably different from that of the Kernohan system (8). It utilizes the presence or absence four specific histologic features: cytologic atypia, mitoses, vascular proliferation, and necrosis. The histologic grade is derived from the number of positive variables. The variables are not "weighted". Thus, a neoplasm with cytologic atypia, mitoses, vascular proliferation, and necrosis is a grade IV, whereas the lesion with cytologic atypia and mitoses is a grade II. Applied to large numbers of patients at the Mayo Clinic, this system appears to have good interobserver consistency and separates neoplasms into four distinct groups. It is of considerable interest that these variables appear to occur in a step-wise fashion in neoplasms of increasing biological malignancy, that is, beginning with cytologic atypia, followed by mitoses, vascular proliferation, and finally, necrosis. If the few patients with the grade I are excluded, the survival curves of the three remaining groups are extremely close to the three groups of neoplasms in the BTSG.

As is obvious from the above discussion, all grading systems are based on a subjective analysis and, even when the details are specified in print, the systems are difficult to compare. It must also be noted that clinical variables such as age, duration of preoperative symptoms, and degree of neurologic deficit are also powerful influences on survival and unless data concerning these are provided in a scientific report, and they often are not, it becomes very difficult to compare survival curves and make inferences as to the effect of different grading systems or treatment (7,9,10). It is hoped that more agreement can be reached on grading systems or, at least, that all reports contain a precise reference by whom and how neoplasms are graded.

It is obvious that more objective means of grading are desirable to supplement or replace the present systems. Two immunohistochemical means to document proliferating or cycling cells are available and under active investigation.

The technique which has received the most study is that utilizing the monoclonal antibody to bromodeoxyuridine (BrdU) (11). The latter compound is injected into the patient at the time of surgery and is detected in paraffin-embedded tissues by a monoclonal antibody. The nuclei of cells in the DNA S-phase of the cell cycle are stained. This makes this technique equivalent in many respects to the classic tritiated thymidine autoradiography, but without the need for a radioisotope and the somewhat more complicated method for autoradiography. These studies have demonstrated a generally close relationship between the classic histologic grade of gliomas and suggest that this technique could be useful in prognostication.

Another approach is application of the monoclonal antibody Ki-67 to detect a nuclear antigen expressed in cycling cells, that is, all phases of cell cycle except G0 (12). This technique does not require the intraoperative administration of a compound, but must be applied to frozen sections. A detailed comparisons between Ki-67 and BrdU are not yet available although, theoretically, more cells should be stained with the Ki-67 method than that of the BrdU since Ki-67 is expressed in more phases of the cell cycle. Preliminary data would suggest that there is a general correlation between the staining index of Ki-67 and the classic histologic grade. Both the studies of Ki-67 and BrdU note that certain lesions in the astrocytoma or anaplastic astrocytoma groups have a particular high staining index. This may suggest that in these lesions the histologic appearances understate the malignancy of the neoplasm. If these neoplasms behave in an aggressive fashion, then it would appear that these immunodiagnostic methods may be a valuable supplement to the classic morphological grading systems in such individuals.

Even newer approaches to the grading of gliomas can be found in studies of gene amplification in gliomas. Several studies have noted the amplification of the erb B oncogene (epidermal growth factor receptor) in somewhat less than half of glioblastomas, and the general absence of amplification in anaplastic astrocytomas and oligodendrogliomas of both well differentiated and anaplastic nature (13). Within the glioblastoma group there were no differences in survivals in patients whose neoplasms were amplified and those whose tumors were not. However, the presence of amplification in certain "anaplastic astrocytomas" was noteworthy and perhaps the presence of amplification in the histologically grade III neoplasm may indicate that the lesion was in fact a glioblastoma in which areas of necrosis and/or vascular proliferation had not been sampled. Clearly, extensive studies of gene amplification and survival remain to be performed but it seems likely that such approaches will be utilized increasingly in the grading of gliomas.

TOPOGRAPHY OF ASTROCYTOMAS

Prior to the application of computerized tomography (CT) there was considerable pessimism concerning the "localization" of the glioblastoma because of the great rarity of surgical cure. However, by CT the neoplasm appeared as a bright contrast-enhancing rim which generally is discrete from the surrounding brain. This rim is surrounded in turn by an area of variable size and configuration consistent with "edema". This appearance of a "discrete" lesion with "peritumoral" edema suggested that the lesions were "localized". Supporting this concept was a detailed study analyzing the patterns of recurrence following treatment. Since the vast majority of lesions occurred within a margin two centimeters from the edge of contrast enhancement, it was inferred that most of the original tumor cells were confined to this area (15). Although an extension of two centimeters will, in the brain, carry many lesions into vital and irresectable structures, it nevertheless lent encouragement to "local" therapies of this disease. Therapies such as interstitial radiotherapy and intraarterial chemotherapy are based largely on the assumption that most tumor cells lie within a acceptable field of therapy. Our own investigations of untreated glioblastomas studied postmortem suggest that the topographic anatomy of the glioblastoma is a complicated issue and that generalizations about the extent of the neoplasm as determined by CT studies are often incorrect (15). Of 15 cases studied in the whole brain sections, three extended past the two centimeters margin and it seems likely that additional cases would extend in a similar fashion if the lesions had been studied in three dimensions. The studies also suggested that the presence of "cerebral edema" as determined by computerized tomography in some instances vastly overstated the extent of the neoplasm, whereas in other instances the neoplastic cells extended beyond the radiographic area of "edema". For the purposes of intraarterial therapy, the study was of interest since it revealed that of the 15 neoplasms, only three were confined within the distribution of either one internal carotid or one posterior cerebral artery by pathologic study to the distribution of either the anterior or the posterior circulation. Thus, there were few "rules of thumb" detected with the exception that fiber pathways such as the corpus callosum are especially prone to conduct neoplastic cells beyond from the tumor bed.

With its marked sensitivity for detection of water when utilized in the T2 weighted mode, magnetic resonance imaging (MRI) has generated concern that the glioblastoma multiformes are considerably more extensive than expected on the basis of computerized tomographic images. In many cases the "water" surrounding a glioblastoma is very extensive in the ipsilateral hemisphere and can even extend into the contralateral brain. Although many have felt that the MRI often overestimates the extent of a malignant glioma, a study utilizing stereotactic biopsies suggests that this might not be the case. In a series of patients from the Mayo Clinic, comparisons between the histologic findings and magnetic resonance images suggest in the glioblastoma malignant cells extend as far if not further than the abnormal T2 signal (20). In these studies, complete maps could not be done and it remains to be seen whether the neoplastic cells extend as far as the edematous area in all areas and in all cases. It seems likely that MRI can, in some cases, overemphasize the extent of the lesion much as is done by the peritumoral area of low density as seen by computerized tomography. The as yet unresolved topography of malignant gliomas is therefore an increasingly important issue in light of the wide use of local therapies such as interstitial radiotherapy.

REFERENCES

1. Kernohan JW, Mabon RF, Svien HJ, Adson AW (1949) A simplified classification of the gliomas. Proc Staff Meet Mayo Clin 24:71-75
2. Kernohan JW, Sayre GP (1952) Atlas of tumor pathology, Section X, Fascicle 35, Tumors of the central nervous system. Armed Forces Institute of Pathology
3. Nelson JS, Tsukada Y, Schoenfeld D, et al (1983) Necrosis as a prognostic criterion in malignant supratentorial, astrocytic tumors. Cancer 52:550-554
5. Zülch KJ (1979) Histological typing of tumours of the central nervous system. World Health Organization, Geneva, pp 43-45, 50
6. Brücher JM, Dalesio O, Solbu G (1987) In Chatel M, Darcel F, Pecker J (eds) Brain Oncology. Dordrecht, Martinus Nijhoff, pp 237-242
7. Burger PC, Vogel FS, Green SB, Strike TA (1985) Glioblastoma and anaplastic astrocytoma: pathologic criteria and prognostic implications. Cancer 56:1106-1111
8. Daumas-Duport C, Scheithauer B, O'Fallon J, Kelly P Grading of astrocytomas: a simple and reproducible grading method. Cancer (in press)
9. Cohadon F, Aouad N, Rougier A, Vital C, Rivel J, Dartigues JF (1985) Histologic and non-histologic factors correlated with survival time in supratentorial astrocytic tumors. J Neuro-Oncol 3:105-111
10. Burger PC, Green SB (1987) Patient age, histologic features, and length of survival in patients with glioblastoma multiforme. Cancer 59:1617-1625
11. Hoshino T, Nagashima T, Murovic JA, Wilson CB, Edwards MSB, Gutin PH, Davis RL, DeArmond S J(1986) In situ kinetics studies on human neuroectodermal tumors with bromodeoxyuridine labeling. J Neurosurg 64:453-459
12. Burger PC, Shibata T, Kleihues P (1986) The use of monoclonal antibody Ki-67 in the identification of proliferating cells: application to surgical neuropathology. Am J Surg Pathol 10:611-617
13. Bigner SH, Burger PC, Wong AJ, Werner MH, Hamilton SR, Muhlbaier LH, Vogelstein B, Bigner DD (1988) Gene amplification in malignant human gliomas: clinical and histopathologic aspects. J Neuropathol Exp Neurol 47:191-205
14. Hochberg FH, Pruitt A (1980) Assumptions in the radiotherapy of teh glioblastoma multiforme. Neurology 30:907-911
15. Burger PC, Shibata T, Kleihues P (1988) Topographic anatomy and CT correlations in the untreated glioblastoma multiforme. J Neurosurg 68:698-704
16. Kelly PJ, Daumas-Duport C, Kispert DB, Kall BA, Scheithauer BW, Illig JJ (1987) Image-based stereotaxic serial biopsies in untreated intracranial neoplasms. J Neurosurg 66:865-874

Cerebral gliomas. G. Broggi and M.A. Gerosa, editors

CYTOGENETICS OF HUMAN GLIOMAS

ROSARIO CASALONE
Istituto di Biologia Generale e Genetica Medica, Università di Pavia, Laboratorio di Analisi, Ospedale di Circolo, Varese, C.P. 217, I 27100 Pavia, Italy.

INTRODUCTION

With the introduction of the chromosome banding techniques and the improvement of ceel culture methods, "non random" clonal chromosome abnormalities have been exactly defined in different tumors.

A vast selection of information is available on chromosomal changes in Leukemia, where specific markers of various leukemic entities, (ex. translocation 9/22 in Chronic Myelocytic Leukemia; translocation 8/21 in Acute Myelocytic Leukemia M_2-FAB etc.) are of relevance in the diagnosis and prognostic evaluation of these diseases [1].

The biologic significance of these cytogenetical results has been confirmed by the appearance of human cellular oncogenes located on the same chromosomal bands involved in the breakpoints that lead to the specific chromosome rearrangements in different tumors [2].

The cytogenetic situation with solid tumors has, until recently, lagged behind that of the Leukemias, for technical reasons and for the difficulties in the interpretation of the more complex karyotypes.

Among the neurogenic tumors, the most reliable results have been established in Retinoblastoma and Meningioma. In this latter monosomy and rearrangements of chromosome 22 are specific and simple clonal abnormalities [3].

Regarding the gliomas the situation is, from a cytogenetical point of view, more complex because of the relatively few cases analyzed and their cytogenetic heterogeneity [4, 7].

MATERIAL AND METHODS

Collection of tumor specimen

Dissected pieces of nonnecrotic tumors were aseptically obtained during surgery, and immediately delivered to the cytogenetic laboratory in sterile tubes containing growth medium (ex RPMI with 20% fetal calf serum and antibiotics).

Tumor dissociation

Mechanical: tumor tissues placed in petri dishes were cut with scissors into small pieces of 1-2 mm_2 in diameter. The fragments and disaggregated cells were introduced into 25 cm^2 culture flasks with the growth medium and cultured at 37°C in 5% CO_2 until a good number of mitoses were detected at the microscopic observation.

Enzymatic disaggregation: tissue fragments were incubated at 37°C for 1-6 hours, suspended in an enzymatic solution containing 0.02% Collagenase (125 U/mg), 0.05% Pronase (45 PKU/mg B grade) and 0.02% DNase. The cell suspensions obtained were centrifugated, resuspended in 25 cm^2 flasks with growth medium and incubated at 37°C in a 5% CO_2 atmosphere.

Harvesting procedure: cultures were observed daily and chromosomal analysis was performed after 24-48 hours and 7-14 days.

Colcemid is added to the cultures (0.02 g/ml final conc.) for 3 hours; the cells were removed by trypsinization, treated in hypotonic solution (0.075 MKCI) for 10 minutes and fixed in a methanol/acetic acid mixture [3, 1]. QFQ and/or GTG banding techniques were applied [3, 6, 8].

Interpretation of chromosome aberration

For a correct estimation of cytogenetic findings in solid tumors, the following indications should be taken into consideration:

1. Metaphases obtained from direct preparations (24-48 hours of culture) or from short term cultures (7-14 days) and after no more than 2° passage can only be considered.
2. A sufficient number of good quality metaphases is required (10-25) for an exact definition of the numerical and structural rearrangements.

3. Only clonal aberrations may be considered: a clone is defined when at least two metaphases with the same structural aberration (deletion, translocation, etc.), or three metaphases with the same numerical aberration are found. The term "Stem-line" indicates the most frequent chromosome constitution of a tumor cell population; all other lines are termed "Side-lines".
4. The patients constitutional karyotype has to be determined from peripheral lymphocytes, as a control.

CYTOGENETICAL FINDINGS (A REVIEW)

There are a few reports on karyotypes of malignant human gliomas (malignant astrocytic tumors including anaplastic astrocytoma, glioblastoma multiforme and gliosarcoma).

The largest collection is that of Mark [4] who demonstrated that 37 of 50 gliomas had near-diploid stemlines and that 26% contained double minutes (DM). This study was, however, performed prior to the banding era. Other cases were successfully reported [5, 6, 7].

In our review, we considered only the recent statistics reported by Bigner et al. [6] and Rey et al. [7], which followed the conditions discussed in the previous paragraph.

Clonal chromosome abnormalities were found in 56 of 61 cases analyzed (92%). Normal karyotypes were found in five cases; in these tumors (without any abnormalities), normal diploid cells would, however, represent nontumoral elements. Two thirds of the tumors were karyotypically homogeneous (only one abnormal cell line), whereas one third was heterogeneous (stem line plus different side lines). This situation may reflect different histopathological features.

Numerical anomalies: the most consistent and probably earliest chromosomal abnormalities of gliomas consist of grains or losses of entire chromosomes. Numerical anomalies were present in the totality of abnormal cases.

In 90% of the cases with numerical deviations, the karyotype is near-diploid (45-47 chromosomes) and in 25% of the 56 abnormal cases, a diploid normal cell line is also present.

Only 10% of the cases show a tri-tetraploid karyotype (69-92 chromosomes). Numerical anomalies were present as only abnormalities in the 32% of the abnormal cases; the remaining cases are associated with structural deviations.

In conclusion, the majority of the cases show simple numerical abnormalities well defined, with a near diploid chromosome number and often associated with a complete normal diploid line.

It confirms our statistic in which 5 of 8 cases of malignant gliomas showed clonal anomalies (3 cases loss of Y, one case loss of Y and 22, one case monosomy 17).

The frequency of involvement of individual chromosomes in clonal numerical aberrations in malignant gliomas (61 cases) is reported in Fig. 1.

As can be seen, chromosome 7 is the most frequently gained (in contrast, no cases with monosomy 7 were found), followed by chromosome 19, 20 and 3.

Chromosome 10 is the most frequently lost (no cases with trisomy 10 were found), followed by chromosome Y, 22, 14, 17 and 9.

It clearly shows that gains of 7 and losses of 10 (always associated) are characteristic of gliomas, but the biological significance of these changes remains speculative.

The erb-b oncogene has been localized to the chromosome 7 (7p12-14) [9] and the locus encoding for Epidermal Growth Factor Receptor (EGFR) has been mapped to 7p13 - 7p11 [9].

The presence of polysomy 7, however, is probably unrelated to amplification of the EGFR gene [10].

The role of loss of chromosome 10 is unknown; no oncogenes have been localized in fact to the chromosome 10.

Finally, the striking association between gains of 7 and losses of 10, suggests that an integrative relationship between genes located on these two chromosomes may exist.

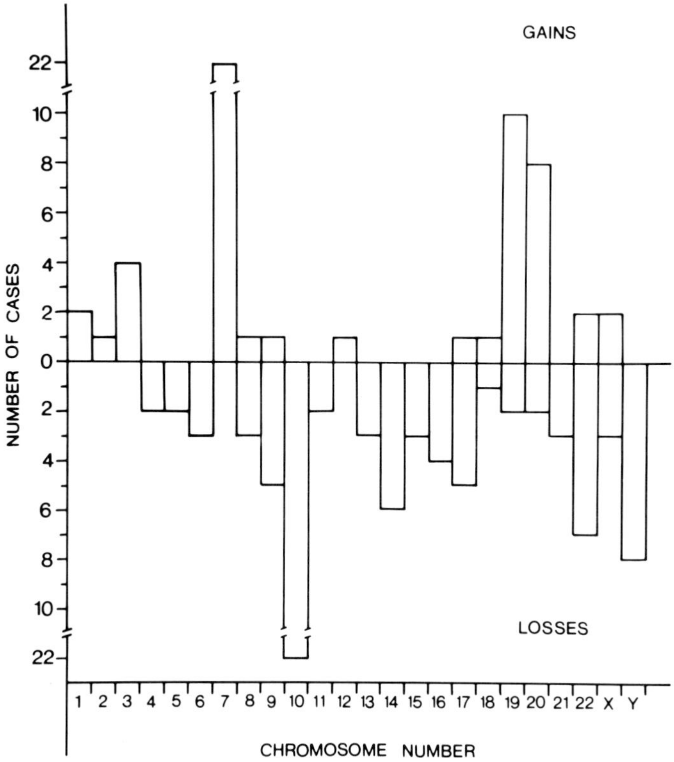

Fig. 1
Histogram showing the frequency of involvement of individual chromosomes in clonal numerical aberrations in gliomas.

Structural anomalies: the most specific structural anomaly of the gliomas is double minutes (DM), acentric fragments, present in 30% of the 61 cases studied. DMS are generally considered genes (oncogenes or genes responsible for drug resistance) [11].

It is unlikely that these amplified genes are related to drug resistance, as many of the patients studied received chemotherapy.

In contrast, a correlation between the presence of DMS and amplification of the EGFR genes has been documented [10].

Numerous other structural abnormalities, however, are frequently present in malignant gliomas. In an attempt to map the breakpoints preferentially involved in acquired chromosome abnormalities, we checked all the cases with clonal structural changes among the 61 cases considered. The results are shown in Fig. 2.

The short arm of chromosome 9, in particular the band p13, is the region most frequently involved. This observation suggests that a group of genes, important in glioma evolution, might be located in this region. However, no oncogenes have been located as yet to the short arm of chromosome 9.

From a prognostic point of view, it was proposed that the patients whose tumors revealed no aberrations or aberrations restricted to loss of sex chromosomes, showed prolonged survival time; on the contrary, a poorer prognosis was found in patients whose tumors showed both numerical and structural anomalies [13]. We think that further studies need to clarify this aspect.

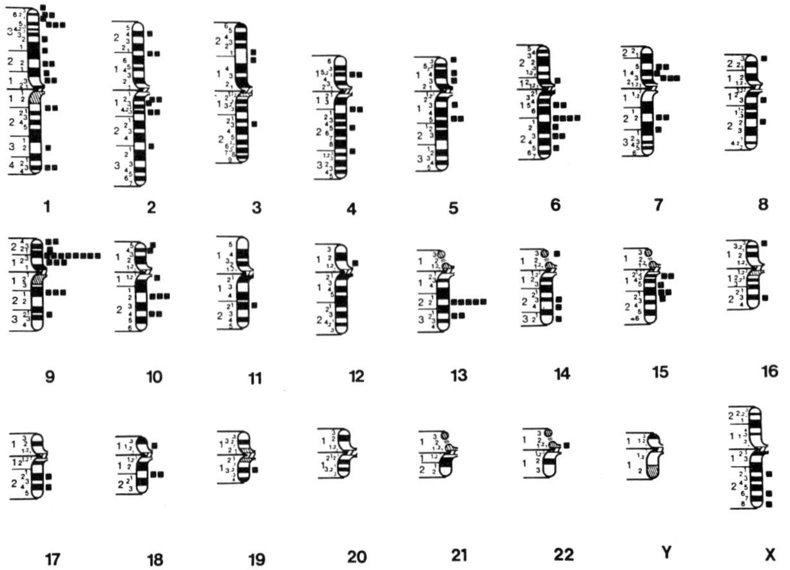

Fig 2.
Location of breakpoints in clonal karyotypic rearrangements in gliomas.

In conclusion, the karyotypic profile of human gliomas is not simple and correlations between cytogenetical aspects and histopathological or clinical patterns are difficult to define, as yet. However, the karyotpye clearly shows a pattern of nonrandom abnormalities; this observation suggests that a set of genes restricted to certain chromosomal regions (chromosome 7, chromosome 10, short arm of chromosome 9, DMS) may be important in the pathogenesis of these tumors.

REFERENCES

1. Sandberg AA (1980) The Chromosomes in Human Cancer and Leukemia, Elsevier/ North Holland, NY.
2. Rowley JD (1983) Nature 301 : 290-291
3. Casalone R, Granata P, Simi P, Tarantino E, Butti G, Bonaguidi R, Faggionato F, Knerich R, Solero L (1987) Cancer Genetic Cytogenet 27 : 145-159
4. Mark J (1971) Hereditas 68 : 61-100
5. Shapiro JR, Yung WKA, Shapiro W (1981) Cancer Research 41 : 2349-2357
6. Bigner SH, Mark J, Bullard D, Mahaley MS, Bigner DD (1986) Cancer Genet Cytogenet 22 : 121-135
7. Rey JA, Bello YM, De Campos JM, Kusak ME, Ramos C, Benitez J (1987) Cancer Genet Cytogenet 29 : 201-221
8. Trent J, Crickard K, Gibas Z, Goodacre A, Pathok S, Sandberg AA, Thompson F, Whang-Peng J, Wolman S (1986) Cancer Genet Cytogenet 19 : 57-66
9. Human Gene Mapping (1985) Cytogenetics Cell Genetics 40 : Nos 1-4
10. Bigner SH, Wong AY, Mark J, Muhlbaier LH, Kinzler KW, Vogelstein B, Bigner D (1987) Cancer Genet Cytogenet 29 : 165-170
11. Gebhart E, Bruderlein S, Tulusan AH, Maillot KV, Bizkmann J (1984) Intl. J. Cancer 34 : 369-373
12. Rey JA, Bello MY, De Campos YM, Kuask ME, Moreno S (1987) Cancer Genet Cytogenet 29 : 223-237
13. Al Saadi A, Latimor F (1985) 23rd Annual Somatic Cell Genetic/San Diego CA.

Cerebral gliomas. G. Broggi and M.A. Gerosa, editors

HUMAN PAPOVAVIRUS BK AND HUMAN CEREBRAL TUMORS

ALFREDO CORALLINI, MONICA PAGNANI, MASSIMO NEGRINI, CLAUDIO MANTOVANI and GIUSEPPE BARBANTI-BRODANO

Institute of Microbiology, School of Medicine, University of Ferrara, Via Luigi Borsari 46, I-44100 Ferrara (Italy)

INTRODUCTION

BK virus (BKV) is a human papovavirus first isolated from the urine of a renal transplant recipient (1). Primary infection generally occurs during childhood and is followed by a persistent, latent infection which is reactivated under conditions of impaired immunological response (2). Serological studies showed that BKV infection is common in normal human population since specific antibodies are found in nearly 80% of adults (3-6). BKV is highly oncogenic in rodents when inoculated either intracerebrally or intravenously in immunosuppressed or immunocompetent hamsters: 73 to 88% of the animals develop tumors. The most frequent tumors belong to 3 histotypes: ependymomas and choroid plexus papillomas, tumors of pancreatic islets and osteosarcomas (7-12). In vitro, BKV and BKV DNA transform hamster, rat, rabbit, mouse, monkey and human cells (13-17). BKV DNA is mostly integrated into cellular DNA in transformed rodent cells (18), while is detected only in a free, episomal state in transformed human cells (13, 14, 17). Induction of brain tumors in rodents and transformation of human cells in vitro raise the possibility of the involvement of BKV in the etiopathogenesis of human brain tumors. We have analyzed primary human brain tumors for the presence of BKV specific DNA, RNA and T antigen.

MATERIAL AND METHODS

Cellular DNA was extracted from human tissues as described previously (19). During the extraction procedure, care was taken to avoid ethanol precipitation in order to save even small amounts of free viral DNA. In each experiment human DNA (20 μg per sample), digested with a restriction endonuclease, was separated in 0,8% (w/v) agarose gels and transferred to nitrocellulose membranes (Schleicher and Schuell BA85 , Dassel, FRG) according to Southern (20). Blots were hybridized and washed as described (21). The dried membranes were exposed to a Kodak X-Omat SO-282 film at -70°C with an intensifying screen. Tumor RNA, extracted with guanidinium isothiocyanate, was spotted on nitrocellulose filters (20 μg per sample) and hybridized as decribed by Thomas (22). BKV T antigen was assayed in tumor tissues by a solid phase enzyme-linked immunoadsorbent assay (ELISA) (23). BKV-transformed hamster cells, 100% positive for T antigen by immunofluorescence, were used as a positive control in ELISA tests. Human embryonic fibroblasts (W138 cells), permissive to BKV lytic infection, were transfected with 40 μg of total tumor DNA by the calcium phosphate precipitation technique (24) as modified by Wigler et al. (25). DNA sequencing was performed by the dideoxy-nucleotide chain termination method (26), using M13 mp18 and 19 as cloning vectors (27).

RESULTS

The results of the analysis of BKV DNA sequences in brain tumors by Southern blot hybridization are presented in Table I. Nineteen out of 74 (25,6%) human tumors were positive for BKV DNA, whereas tumors of other sites or normal tissues were negative. No specific association of BKV to any brain tumor hystotype was found, except in the case of glioblastoma where 50% of the tumors contained BKV DNA.

TABLE I

BKV DNA IN HUMAN NEOPLASTIC AND NORMAL TISSUES

Brain tumors	Positive samples/ samples analysed (%)	Rescue of BKV DNA/ positive DNAs tested
Spongioblastoma	1/3	1/1
Glioblastoma	9/18	1/9
Astrocytoma	2/11	1/2
Meningioma	2/20	1/3
Ependymoma	1/3	1/2
Oligodendroglioma	1/1	1/1
Neurinoma	1/6	
Medulloblastoma	0/1	
Undifferentiated tumor	0/2	
Adenoma of hypophysis	1/1	
Rhabdomyosarcoma	0/1	
Lymphoma	0/1	
Metastasis of adenocarcinoma	1/4	
Metastasis of sarcoma	0/2	
Total	19/74 (25.6)	6/18
Other tumors*	0/10	
Normal tissues**	0/15	

* 2 cervical, 2 vulvar, 3 labial and 3 kidney carcinomas.
** Tissue samples removed surgically or at biopsy from living patients: skin, brain, pancreas, uterine cervix, peripheral blood leukocytes. Tissue taken at autopsy: brain.

In all the positive samples, viral DNA was free, in an episomal state and in a low copy number: 0,2 to 2 genome equivalents per cell (Fig. 1). No viral DNA sequences integrated into cellular DNA were detected.

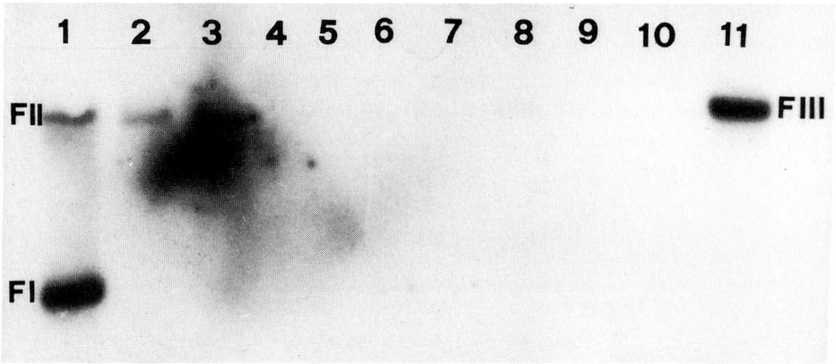

Fig. 1. Southern blot hybridization of BamHI-digested human tumor DNAs to a ^{32}P-labeled BKV DNA probe. Lane 1 and lane 11 contain 2.5 genome equivalents of BKV DNA form I (FI) and form II (FII) or form III (FIII, full length linear form obtained by cleavage of BKV DNA with BamHI). Lanes 2-10 contain DNA from human brain tumors (20 µg per lane).

Total RNA analyzed by dot blot hybridization with a ^{32}P-labelled BKV DNA probe revealed the presence of BKV-specific RNA in 11 out of 26 tumors. BKV T antigen was assayed in 8 human tumors by ELISA: 5 glioblastomas, including a relapse, and a primary meningioma were positive, whereas another meningioma and an ependymoma were negative. Positive and negative results for BKV DNA, RNA and T antigen are concordant in all the tumors analyzed except for a meningioma that is negative for DNA and RNA and positive for T antigen (Table II).

Since BKV DNA is in a free state in all the positive tumors, virus rescue was attempted by transfection of tumors DNAs into WI38 cells, permissive to BKV lytic infection. From 6 out of 18 tumor DNAs a virus was recovered (Table I) with the biological and antigenic properties of BKV. Restriction enzyme analysis of viral DNAs shows that the genomes of all the rescued viruses are similar to each other, but different from the genome of BKV wild type (BKV-WT)(Fig. 2).

TABLE II

BKV DNA, RNA AND T ANTIGEN IN HUMAN TUMORS

Tumor histotype	No. of samples	DNA	RNA	T antigen
Glioblastoma	4	+	+	+
Glioblastoma (relapse)	1	+	+	+
Meningioma	1	-	-	+
Meningioma	2	-	-	
Meningioma	1	-		-
Ependymoma	1	+	+	
Ependymoma	1	-	-	
Ependymoma	1	-		-
Astrocytoma	1	-	-	
L578*	1	+	+	+
HKBK-DNA-4*	1	+	+	+
Normal Brains	5	-	-	-

* Hamster cells transformed by BKV DNA, 100% positive for T antigen by immunofluorescence.

Digestion of BKV-RL and BKV-E DNAs with EcoRI, which cleaves BKV DNA once, produces full-length linear DNA form III (FIII) migrating slightly ahead of BKV-WT DNA FIII (Fig. 2, lanes 1-3). Digestion of DNA from the two variants with PvuII yields two fragments: the larger one migrates faster than the corresponding PvuII fragment of BKV-WT DNA (Fig. 2, lanes 7-9). BKV-RL and BKV-E DNAs, cleaved with HindIII, produce only three fragments, instead of the four obtained with BKV-WT DNA: fragment A, which comigrates with the corresponding fragment A of BKV-WT DNA, and two additional fragments B and C, whereas fragment D is absent (Fig. 2, lanes 4-6). This restriction pattern is consistent with an insertion in the region of the genome corresponding to HindIII fragment C and with a deletion involving a portion of HindIII fragments B and D. As a consequence of the deletion, the HindIII site between fragments B and D of BKV-WT is lost, producing a fused fragment larger than fragment B.

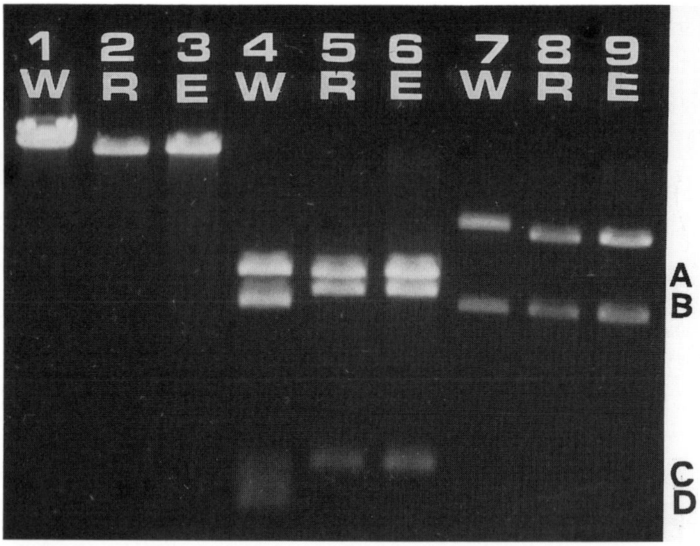

Fig. 2. Agarose gel electrophoresis of DNAs from BKV-WT (W) and BKV variants BKV-RL (R) and BKV-E (E). The variants were rescued respectively from an astrocytoma and an ependymoma. Viral DNAs were cleaved with EcoRI (lanes 1-3), HindIII (lanes 4-6) and PvuII (lanes 7-9). A, B, C and D indicate the four HindIII fragments of BKV-WT DNA.

These results were confirmed by nucleotide sequence analysis of BKV-E DNA, showing an insertion of 80 nucleotides in HindIII fragment C and a deletion of 253 nucleotides between HindIII fragments B and D.

DISCUSSION

These data provide evidence for a specific association of BKV with some human brain tumors. In fact, even if samples of normal brain may be insufficient to rule out a BKV latent infection in the brain, in two studies (28, 29) performed to investigate the sites of BKV latency in human population, BKV DNA was not found in the brain, but it was detected by others in several normal human tissues (30). It is notable that viral DNA in human tumors is in an episomal state like in human cells transformed in vitro by BKV (13, 14, 17). The presence of BKV-specific RNA and T antigen in some brain tumors indicates that the viral genome is expressed. Since T antigen is the primary effector of the oncogenic properties of papovaviruses (31), the expression of BKV T antigen in these tumors may indicate an effective role of BKV in the process of carcinogenesis. All viruses rescued from brain tumor DNAs are similar to each other and to BKV-IR, a BKV variant previously isolated from a human insulinoma (32); however they differ from BKV-WT, raising the possibility that a specific BKV variant is associated with human tumors. Nucleotide sequence analysis of variants BKV-IR and BKV-E showed that the inserted nucleotides can form an insertion sequence (IS)-like structure. The IS-like element contains in its loop two enhancer repeats. This hypothetical transposable element might be involved in the neoplastic process by inserting into the cell genome, thereby inducing a mutagenic effect or activating the expression of cellular oncogenes.

ACKNOWLEDGEMENTS

We thank Mr. A. Bevilacqua for excellent technical assistance and for typing the manuscript. This research was supported by grants from the Italian National Research Council (Special Project "Oncology" contract 87.01173.44) and from Associazione Italiana per la Ricerca sul Cancro (A.I.R.C.).

REFERENCES

1. Gardner SD, Field AM, Coleman DV, Hulme B (1971) Lancet 1:1253-1257

2. Padgett BL, Walker DL (1976) Prog Med Virol 22:1-35

3. Gardner SD (1973) Brit Med J 1:77-78

4. Shah KV, Daniel RW, Warszawski RM (1973) J Infec Dis 128:784-787

5. Portolani M, Marzocchi A, Barbanti-Brodano G, La Placa M (1974) J Med Microbiol 7:543-546

6. Brown P, Tsai T, Gajdusek DC (1975) Amer J Epidemiol 12:331-340

7. Costa T, Yee C, Tralka TS, Rabson AS (1976) J Nat Cancer Inst 56:863-864

8. Uchida S, Watanabe S, Aizawa T, Kato K, Furuno A, Muto T (1976) Gann 67:857-865

9. Corallini A, Barbanti-Brodano G, Bortoloni W, Nenci I, Cassai E, Tampieri M, Portolani M, Borgatti M (1977) J Nat Cancer Inst 59:1561-1563

10. Corallini A, Altavilla G, Cecchetti MG, Fabris G, Grossi MP, Balboni PG, Lanza G, Barbanti-Brodano G (1978) J Nat Cancer Inst 61:875-883

11. Uchida S, Watanabe S, Aizawa T, Furuno A, Muto T (1979) J Nat Cancer Inst 63:119-126

12. Corallini A, Altavilla G, Carrà L, Grossi MP, Federspil G, Caputo A, Negrini M, Barbanti-Brodano G (1982) Arch Virol 73:243-253

13. Purchio AF, Fareed GC (1979) J Virol 29:763-769

14. Takemoto KK, Linke H, Miyamura T, Fareed GC (1979) J Virol 29:1177-1185

15. Howley PM (1980) In: Klein G (ed) Viral Oncology. Raven Press, New York, pp 489-550

16. Padgett B (1980) In: Tooze J (ed) Molecular Biology of Tumor Viruses. II. DNA Tumor Viruses. Cold Spring Harbor Laboratory, New York, pp 339-370

17. Grossi MP, Caputo A, Meneguzzi G, Corallini A, Carrà L, Portolani M, Borgatti M, Milanesi G, Barbanti-Brodano G (1982) J Gen Virol 63:393-403

18. Meneguzzi G, Chenciner N, Corallini A, Grossi MP, Barbanti-Brodano G, Milanesi G (1981) Virology 111:139-153

19. Chenciner N, Meneguzzi G, Corallini A, Grossi MP, Grassi P, Barbanti-Brodano G, Milanesi G (1980) Proc Natl Acad Sci USA 77:975-979

20. Southern EM (1975) J Mol Biol 98:503-517

21. Chenciner N, Grossi MP, Meneguzzi G, Corallini A, Manservigi R, Barbanti-Brodano G, Milanesi G (1980) Virology 103:138-148

22. Thomas PS (1980) Proc Natl Acad Sci USA 77:5201-5206

23. Corallini A, Pagnani M, Viadana P, Silini E, Mottes M, Milanesi G, Gerna G, Vettor R, Trapella G, Silvani V, Gaist G, Barbanti-Brodano G (1987) Int J Cancer 39:60-67

24. Graham FL, van der Eb AJ (1973) Virology 52:456-467

25. Wigler M, Pellicer A, Silverstein S, Axel R, Urlaub G, Chasin L (1979) Proc Natl Acad Sci USA 76:1373-1376

26. Sanger F, Nicklen S, Coulson AR (1977) Proc Natl Acad Sci USA 74:5463-5467

27. Yanish-Perron C, Vieiria J, Messing J (1985) Gene 33:103-119

28. Chester PM, Heritage J, McCance DJ (1983) J Infect Dis 147:476-484

29. Grinnell BV, Padgett BL, Walker DL (1983) J Infect Dis 147:669-675

30. Pater MM, Pater A, Fiori M, Slota J, Di Mayorca G (1980) In: Essex M, Todaro G, zur Hausen H (eds) Viruses in Naturally Occurring Cancers. Cold Spring Harbor Laboratory, New York, pp 329-341

31. Tegtmeyer P (1980) In: Tooze J (ed) Molecualr Biology of Tumor Viruses II: DNA Tumor Viruses. Cold Spring Harbor Laboratory, New York, pp 297-337

32. Caputo A, Corallini A, Grossi MP, Carrà L, Balboni PG, Negrini M, Milanesi G, Federspil G, Barbanti-Brodano G (1983) J Med Virol 12:37-49

ONCOGENE EXPRESSION IN HUMAN BRAIN TUMORS

GIULIANO DELLA VALLE[1], DANIELA TALARICO[2], CATERINA FOGNANI[2], ANTONIO F. PEVERALI[2], ELENA RAIMONDI[2], LUIGI DE CARLI[2] and MASSIMO GEROSA[3]
[1]Dip. di Biologia Animale, Università di Catania, via Androne 81, 95124 Catania, [2]Dip. di Genetica e Microbiologia, Università di Pavia and [3]Dip. di Neurochirurgia, Università di Verona, Italy

INTRODUCTION

Cellular oncogenes may acquire oncogenic potential through genetic events leading to: (i) alteration of the coding sequence, resulting in a gene product with aberrant activity; (ii) increased expression of the normal allele, caused by gene amplification or loss of control of its transcriptional activity (1).

Transcriptional activation of cellular oncogenes has been demonstrated in several malignant solid tumors, as well as in human malignant cell lines (2). In the case of neoplasias of neuro-ectodermal origin, particularly brain tumors, a key role in neoplastic progression has frequently been attributed to abnormal expression of the epidermal growth factor (EGF) receptor (3). In human glioblastomas high levels of EGF receptors are commonly associated with increased number of EGF receptor gene due to gene amplification or abnormal chromosome segregation (4-6). Indeed, extra copies of chromosome 7, where the EGF receptor gene has been localized (7), is a frequent finding in glioblastoma derived cell lines (8).

The EGF receptor gene is the cellular homologue of the v-*erb*-B oncogene of the avian erythroblastosis virus. The protein coded by v-*erb*-B is closely related to the transmembrane and to the catalytic intracellular domains of the EGF receptor (9). As with several other growth factor receptors, the activation of the EGF receptor triggers a complex array of intracellular responses, including induction of an intrinsic tyrosine kinase activity. In competent cells, this complex chain of events results in DNA synthesis and, ultimately, in cell division (10).

Human *ras* oncogenes (Ha-*ras*-1, Ki-*ras*-2 and N-*ras*) can lead to the malignant transformation of non-tumorigenic cells by either single point mutation or aberrant expression of the unaltered proto-oncogenes (1, 11). Enhanced expression of ras genes, resulting from gene amplification, has been observed in various tumours (12), and high levels of ras protein have been correlated with the depth of colon carcinoma invasion (13). Human ras oncogenes encode 21-KD proteins (p21) which are membrane-associated guanosine triphosphate-binding (GTP) proteins. The proteins have an intrinsic low GTPase activity that, in certain cases, is impaired in the mutated form (11).

Two recent experimental evidences suggest that ras proteins may play a role in the EGF receptor system: I, the growth promoting effect of EGF is enhanced in mouse fibroblasts which express high levels $p21^{N\text{-}ras}$ (14); II, in membranes isolated from ras-transformed rat cells the guanine nucleotide binding activity of $p21^{ras}$ is stimulated by EGF (15). We therefore analysed a series of human malignant brain tumour cell lines and surgical biopsies to investigate the possibility that in glioblastomas increased levels of EGF receptor might be associated with ras gene overexpression.

MATERIALS AND METHODS

Cell lines

Five human glioblastoma multiforme-derived primary cell lines (Hu70, Hu104, Hu112, Hu195 and Hu197) have been established from surgical specimens according to procedures previously described (16). Monolayer cultures were grown in Ham's F12 supplemented with 20% fetal calf serum (Flow Laboratories), incubated in a 5% CO_2 humidified atmosphere and split 1:2-1:4 when confluent. Every 5-7 culture passages, the cells were tested for the expression of the astrocytic markers (GFAP), for phenotypic characterization, as previously reported (17).

Long-term cultures of non-neoplastic glial cells (Hu175) served as controls.

Surgical specimens

Primary brain tumours and control brain tissue samples were obtained, either by endocranial biopsy or during major neurosugery, from untreated patients.

All specimens were frozen immediately after excision and stored in liquid nitrogen.

Nucleic acid analysis

Culture dishes were washed twice with PBS, cells were scraped and collected by centrifugation. Frozen tissue samples were pulverized in liquid nitrogen. DNA was extracted by standard procedures which include SDS lysis, proteinase K treatment, phenol-chloroform extraction, RNAse digestion and ethanol precipitation.

DNAs, quantitated by A_{260}, were serially diluted to 100μg, 50μg and 25μg per ml and denatured in 0,4 M NaOH for 15m at room temperature. The solutions were then brought to 1M ammonium acetate and aliquots (100μl) containing 4μg, 2μg and 1μg were spotted in duplicate onto Amersham's Hybond nylon membranes, presoaked in 0,3 M sodium citrate, 3M sodium chloride (20xSSC). This procedure was performed utilizing a Minifold II apparatus (Schleicher and Schuell). The filters were dried and baked for 3 h at 80°C.

RNA was isolated from cells and frozen tissues by the guanidine thiocyanate/CsCl method described by Chirgwin et al. (18). Dot blot RNA analysis of serially diluted samples (10 μg to 1,25 μg) was performed as described by White and Bancroft (19); samples were dotted onto nylon membranes under slow vacuum suction.

Northern blot analysis and filter hybridizations were performed as previously described (6).

The 1.8 kb *Eco*RI fragment of human EGF receptor cDNA isolated from plasmid pHER-A64-1, the 1.5 Kb *Eco*RI N-*ras* fragment from N-*ras* plasmid (Oncor) and the 0.6 Kb *Bam*HI-*Eco*RI β-actin fragment from pHFβA-3'UT plasmid were used as probes.

RESULTS

EGF receptor gene expression in cell lines and tumors

We estimated the copy number and the expression of EGF receptor gene in five glioblastoma-derived cell lines, in one astrocytoma and four glioblastoma tissue samples (Table I). DNA slot blot analysis showed the presence of extra copies of EGF receptor gene in four out of five cell lines and in two glioblastomas.

We have previously reported that in these cell lines increased dosage of EGF receptor gene was mainly due to the presence of supernumerary copies of chromosome 7, where the gene has been mapped. On the other hand gene amplification should have occurred in HuGl-9 tumor which carried 20-40 gene copies per cell.

RNA analysis showed overexpression of EGF receptor gene in most of the neoplastic cell lines and tumors. The levels of specific mRNAs detected in each line or tumor were correlated with the gene copy number per cell.

TABLE I

EGF receptor gene copy number and mRNA in glioblastoma-derived cell lines and tumors

Cell line or tissue	Origin	Gene copy number / cell[a]	mRNA[b]
Cell line			
Hu70	Glioblastoma multiforme	3-4	+ + +
Hu104	Glioblastoma multiforme	4	+ + + +
Hu112	Glioblastoma multiforme	2	+
Hu195	Glioblastoma multiforme	4	+ +
Hu197	Glioblastoma multiforme	4-5	+ + + +
Hu175	Normal brain	2	+
Tissue			
HuGl-1	Glioblastoma multiforme	2	+ + +
HuGl-3	Glioblastoma multiforme	2	+ + +
HuGl-4	Glioblastoma multiforme	3-5	+ + + +
HuGl-7	Astrocytoma	2	+ +
HuGl-9	Glioblastoma multiforme	20-40	*[c]
Gl-4-Nat	Normal tissue adjacent to HuGl-4	2	+
Nb	Normal brain	2	+

[a]DNA extracted from cell or tissue samples was slot blotted onto nylon membrane and hybridized with EGF receptor cDNA probe. The results are expressed as the amount of labelled probe annealed to the cellular DNA, normalized to that of Hu175 or normal brain (which were taken as 2).

[b]The lavels of mRNA were scored by eye on a relative basis from + to ++++ depending upon the intensity of the autoradiographic signal.

[c]10-30 fold overexpression.

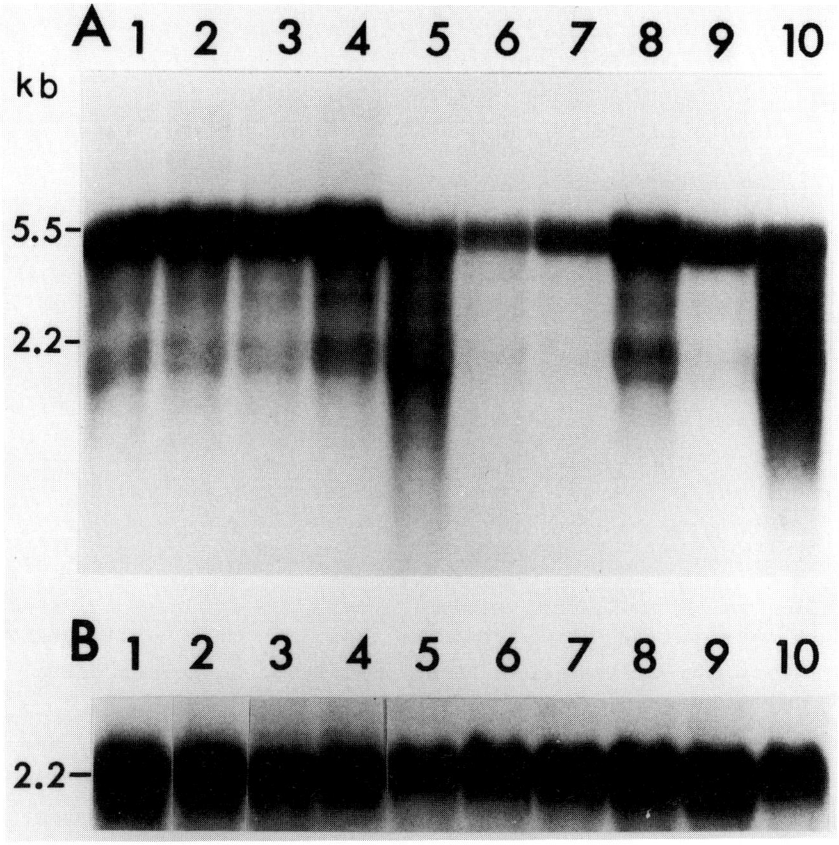

Fig. 1. Expression of N-ras oncogene in glioblastoma derived cell lines and tumors.
Total RNA extracted from glioblastoma cell lines Hu70 (1) , Hu104 (2), Hu112 (3), Hu195 (4), Hu197 (5), normal glial cell line Hu175 (6) and from tissue samples of normal brain (7), glioblastoma multiforme HuGl4 (8), HuGl9 (10), and of astrocytoma HuGl7 (9), were analyzed by Northern blot. The filter was hybridized with radiolabelled N-*ras* specific sequences (A), and, after stripping, re-hybridized with actin specific probe (B). Numbers at the left indicate sizes of major N-*ras* (A) and actin (B) mRNAs.

N-*ras* expression

The expression of N-*ras* oncogene was investigated by Northern blot analysis of the total RNA extracted from the neoplastic cell lines and from tissue samples of three tumors (Fig. 1). RNAs from normal cultured glial cells and normal brain tissue were taken as controls.

All glioblastoma cell lines and tumors showed overexpression (4-8 fold) of N-*ras*. Only a slight increase of specific transcripts was observed in an astrocytoma.

DISCUSSION

We have analyzed the expression of EGF receptor gene and N-*ras* oncogene in glioblastoma cell lines and tumors.

EGF receptor gene was expressed at high level in four cell lines and in all tumor biopsies, compared to normal brain tissue and non-neoplastic glial cells. In most cases the overexpression was related either to amplification or to small increases in number of the EGF receptor gene per cell.

High EGF receptor levels have been reported in several human tumours and tumour-derived cell lines, mainly of neuro-ectodermal origin (3). Enhanced EGF receptor activity has been found in malignant glioma homogenates (3, 4) and in cell cultures (20), but not in normal adult brain (3, 4). An amplified, overexpressed, rearranged EGF receptor has been identified in a human astrocytoma cell line (21), and a high-affinity EGF receptor in a malignant glioma cell line (22). Furthermore EGF receptors seem to play an important role in the physiology of glial cells. It has been reported that EGF can stimulate the growth of mouse astrocytes and normal human glial cells in culture (10).

In four cell lines and in two tumors we observed that overexpression of EGF receptor gene was associated with the presence of high levels of N-*ras* specific transcripts.

Experimental evidence suggests that EGF-r and *ras* proteins, which reside in the same cellular compartment, may ineract directly or indirectly in response to mitogenic stimulation, namely: (i) the growth-promoting effect of EGF is greatly enhanced in NIH 3T3 cells which express high levels of N-*ras* protein (14); (ii) addition of EGF to membranes isolated from c-Ha-*ras*-transformed rat cells enhances the guanine nucleotide binding activity of $p21^{ras}$ in a dose dependent manner (15).

As EGF is present in human cerebrospinal fluid in concentrations comparable to those in plasma (10), it may be postulated that an abundance of EGF receptor and p21$^{N-\underline{ras}}$ may play a key role in glial tumour cell proliferation.

ACKNOWLEDGEMENTS

We wish to thank A. Ullrich for the gift of plasmid pHER-A64-1 and L. Kedes for plasmid pHFβA-3'UT. Our thanks also go to M. Negri for his expert technical assistance and to G. Ranzani for her critical comments on the manuscript.

This work was supported by grants from the Ministero Italiano della Pubblica Istruzione, the Associazione Italiana per la Ricerca sul Cancro (AIRC), the CNR (Progetto Finalizzato Oncologia), and the Regione Veneto. D.T. and A.F.P. are fellows of the AIRC.

REFERENCES

1. Varmus HE (1984) Ann. Rev. Biochem. 18: 553-612.
2. Slamon DJ, de Kernion JB, Verma IM, Cline MJ (1984) Science 224: 256-262.
3. Libermann TA, Razon N, Bartal AD, Yarden Y, Schlessinger J, Soreq H (1984) Cancer Res. 44:753-760.
4. Libermann TA, Nusbaum HR, Razon N, Kris R, Lax I, Soreq H., Whittle N, Waterfield MD, Ullrich A, Schlessinger J (1985) Nature 313:144-147.
5. Henn W, Blin N, Zang KD (1986) Hum. Genet. 74:104-106.
6. Talarico D. Raimondi E, Fognani C, Gerosa MA, Gregotti S, Della Valle G, De Carli L (1987) Cytotechnology 1:41-46.
7. Spurr NK, Solomon E, Jansson M, Sheer D, Goodfellow PN, Bodmer WG, Vennstrom B (1984) Embo J. 3:159-164.
8. Bigner SH, Mark J, Bigner DD. Cancer Genetic. Cytogenet. (1984) 24:163-176.
9. Downward J, Yarden Y, Mayes E, Scrace G, Totty N, Stockwell P, Ullrich A, Schlessinger J, Waterfield MD (1984) Nature 307:321-327.
10. Guroff G (1983) Growth and maturation factors, Vol. I. New York: Wiley.
11. Barbacid MA (1986) A. Rev. Biochem. 56:779-827.
12. Schwab M, Alitalo K, Varmus HE, Bishop JM, George D (1983) Nature 303:497-501.

13. Thor A, Horan Hand P, Wunderlich D, Caruso A, Muraro A, Scholm J (1984) Naure 311:562-565.
14. Wakelam MJO, Davies SA, Houslay MD, McKay I, Marshall CJ, Hall A (1986) Nature 323: 173-176.
15. Kamata T, Feramisco JM (1984) Nature 310:147-149.
16. Rosenblum ML, Gerosa MA, Wilson CB, Barger GR, Pertuiset BF, De Tribolet N, Dougherty DV (1983) J. Neurosurg 58:170-176.
17. Gerosa MA, Rosenblum ML, Tridente G, eds. (1986) Brain tumors: biopathology and therapy. Oxford: Pergamon Press.
18. Chirgwin JM, Przybyla AE, MacDonald RJ, Rutter WJ (1979) Biochemistry 18:5294-5299.
19. White BA, Bancroft FC (1982) J. Biol. Chem. 257:8569-8572.
20. Harsh GR IV, Rosenblum ML, Williams LT (1985) Proc. Am. Ass. Neurol. Sug. :264-265.
21. Filmus J, Pollak MN, Cairncross JG, Buick RN (1985) Biochem. Biophys. Res. Commun. 131:207-215.
22. Westphal M, Harsh GR; Rosenblum ML et al. (in press) Biochem. Biophys. Res. Commun.

© 1989 Elsevier Science Publishers B.V. (Biomedical Division)
Cerebral gliomas. G. Broggi and M.A. Gerosa, editors

CELL KINETICS, AGE AND PROGNOSIS IN MATURE AND ANAPLASIC ASTROCYTOMAS

A. FRANZINI*, G. BROGGI*, C. GIORGI*, L. CAJOLA*, A. ALLEGRANZA**
* Dept. Neurosurgery and ** Dept. Neuropathology, Istituto Neurologico "C. Besta", Milano (Italy)

INTRODUCTION

The fate of supratentorial glial tumors is characterized by soon or late development of malignant behaviour (1, 2, 3). In spite of this dramatic consideration,it is well known that the survival of patients affected by astrocytic tumors may vary between few months and several years depending from the stage of the disease at the time of the diagnosis. Particularly in deep brain glial tumors, the assessment of the prognosis and the choice of treatment are strictly dependent from the histological grading based on the detection of peculiar morphological features in the fragments of neoplastic tissue obtained by serial stereotactic biopsy (4, 5). Nevertheless, in a certain number of patients, the tumor behaviour and the clinical course may be different from the one suggested by histological grading itself; in fact, patients affected by low-grade astrocytoma may undergo fast growth of the tumor in spite of the expected long-lasting stabilized conditions (6, 2). Conversely, some patients belonging to the anaplasic astrocytoma group may present relatively slow growing behaviour. These controversies in the prognostic evaluation of glial tumors motivated the research of objective parameters more predictive of the potential growth of the tumor and more linked to the biology of these lesions. For this purpose, in the past decade, Hoshino et al. introduced the study of the potential proliferative activity of astrocytic tumors by cell kinetics investigations performed in surgically removed neoplasms (2). Nevertheless, in our opinion, the utilization of cell kinetics indices is particularly boosted in the field of stereotactic procedures which include either diagnostic than therapeutical techniques as interstitial irradiation, radiosurgery and stereotactically guided microsurgery (7, 8, 9). Aim of this report is the study of the relationships between cell kinetics indices and other prognostic factors with particular regard to the age of patients.

METHODS AND PATIENTS

Stereotactic procedure. The Riechert frame and apparatus have been utilized. Ventriculography has been performed in stereotactic conditions and the CT images of the lesion have been transposed in the stereotactic planes by mathematical method. The electrical impedance has been monitored along the full transtumoral trajectory to reveal necrotic areas, cysts and major peritumoral edema (4, 6). The Sedan or Nashold bioptic instrument has been utilized to obtain tissue specimens from multiple different targets within the peritumoral areas and the estimated tumoral core. In each patient have been obtained 2-7 tissue samples (average 4 samples). The obtained cylinders of tissue (8 mm long, 1.8 mm diameter) have been longitudinally split to provide mirror specimens for morphological and cell kinetic studies. Finally a small NMR compatible marker has been placed at the deepest target to confirm postoperatively the site of biopsy.

Laboratory procedure. The samples of tissue devoted to cell kinetic studies have been immediately placed in cold RPMI 1640 (GIBCO, Grand Island, NY) medium and processed within 1 hour. The specimens have been incubated in 2 ml of complete medium with 3H thymidine (RPMI 1640 with 20% FCS plus antibiotics - 6 μCi/ml; s.a. 5 Ci/mmole; Radiochemical Centre, Amerham, UK) at 37° C for 1 hour in a shaking water bath. After the incubation, the biopsy specimens have been fixed in Bouin's solution for 1 hour, embedded in paraffin, sectioned at 4 μm for autoradiographic procedure. Deparaffinized slides have been harvested with the stripping film (Kodak AR10, Kodac, Rochester, NY) technique and exposed at 4° C for 10 days. The slides have been developed in Kodac D 19b for 5 minutes at 18° C, fixed and stained with hematoxylin and eosin at 4° C. Finally the slides have been examined by optical microscopy and 1000 to 10000 cells have been scored to count the total number of labelled cells; in labelled cells the number of grains per nucleus resulted always more than 10. The labeling index (LI) has been calculated as the ratio % between labeled cells and total cells.

Patients and evaluation criteria. The patients have been chosen in order to fullfill predetermined criteria for retrospective evaluation. The potential proliferative activity of each tumor has been indicated by the highest LI% value detected along the stereotactic trajectory (6). The selected cases are characterized by the following conditions:

1) deep seated tumors unsuitable of surgical removal and belonging just to mature astrocytoma and anaplasic astrocytoma groups (patients which resulted affected by pilocytic astrocytoma, oligodendroglioma and giant cell astrocytoma have been withdrawn from this study because of the peculiar behaviour of these lesions; patients affected by glioblastoma have not been included in this study because of the poor meaning of cell kinetics investigations in these tumors) (2);
2) availability of clinical and neuroradiological follow-up lasting almost 3 years after the histological diagnosis;
3) No chemotherapeutic or radiotherapeutic treatments performed in the considered periods (3-5 years) after the stereotactic biopsy.

It has to be stressed that in the period considered for this study our strategy in the management of deep seated astrocytic tumors consisted in surveillance for mature astrocytoma and in external radiotherapy for anaplasic astrocytoma. For these reasons the finally selected series included 53 patients affected by untreated mature astrocytoma and 28 patients affected by anaplasic astrocytoma in which radiotherapy has been refused or delayed later than 3 years after the stereotactic biopsy because of familial or geographical reasons.

In the whole considered series the age ranged between 2 and 61 years (mean 32 years; 39 patients were females. The survival, the clinical course and the neuroradiological evolution have been then retrospectively evaluated and correlated to LI and to the age of patients at the moment of stereotactic biopsy.

RESULTS AND CONCLUSIONS

At retrospective evaluation, the outcome and the clinical course in the whole series of patients resulted correlated to the LI% values detected by serial stereotactic biopsy (Fig. 1). Nevertheless, the predictive value of kinetic parameter appears to be much more significative when applied to groups of patients homogeneous as regard to the age and the histological grading. The following remarks concern particularly the relationships between prognosis, age, LI values and histological findings.

a) Tumors in which the LI% resulted higher than 10%, presented soon development of malignant behaviour either in mature than anaplasic astrocytoma series and none of this patients survived longer than 1 year (Fig. 1). This particularly malignant course occurred in 5% of patients belonging to series of mature astrocytoma versus 25% of patients affected by anaplasic astrocytoma. The prognosis in these group of patients has been clearly indicated by LI% values while the histological examination does not suggested go fast and malignant evolution. The higher sensitivity of biological versus morphological investigations in these patients may be due either to the small amount of neoplastic tissue provided by serial stereotactic biopsy for conventional histological examination either to the possible detection of a precox malignant phase not yet histologically evident. The age of patients does not affect the prognosis of these patients with LI values higher than 10%.

b) On the other extremity of the above series are considered tumors in which the LI% values ranged between 0.3% and 5%. Only 11.6% of these patients died during the follow-up period. The mortality in patients affected by anaplasic astrocytoma was 30% versus 5% in patients affected by mature astrocytoma. These data suggest that LI cannot be considered an absolute index of proliferative activity: in fact, similar values may have a different meaning in mature than anaplasic astrocytoma.

c) In the same series of patients characterized by LI below 5%, have been investigated the relationships between age, cell kinetics and survival (Fig. 2). In the group of patients under fourty years old zero mortality occurred in mature astrocytoma and 11.9% mortality in anaplasic astrocytoma series. In patients older than fourty years, the mortality was 20% in mature astrocytoma and 60% in anaplasic astrocytoma. It appears from these data that in group of patients homogeneous as regard to the histological grading and also homogeneous as regard the LI% (below 5%), the prognosis is significatively modulated by the age of patients. The prognostic value of age in glial tumors is well known from statistical analysis (3, 10) and our data suggest that the factors involved are much more linked to the host than to the aggressiveness of the tumors (Fig. 2).

In conclusion, the LI allowed to disclose the amount of tumors candidate to fast malignant evolution either in the group of mature than anaplasic astrocytoma (LI higher than 10%).

In patients with lower LI values the outcome and survival resulted much more favourable. Particularly in the series characterized by LI below 5%. The predictivity of cell kinetic investigations resulted much more accurate when applied to series of patients homogeneous as regard to the age and to the histological grading. In our opinion cell kinetic investigation will boost the accuracy and the indications to diagnostic and therapeutic stereotactic procedures, particularly in young patients.

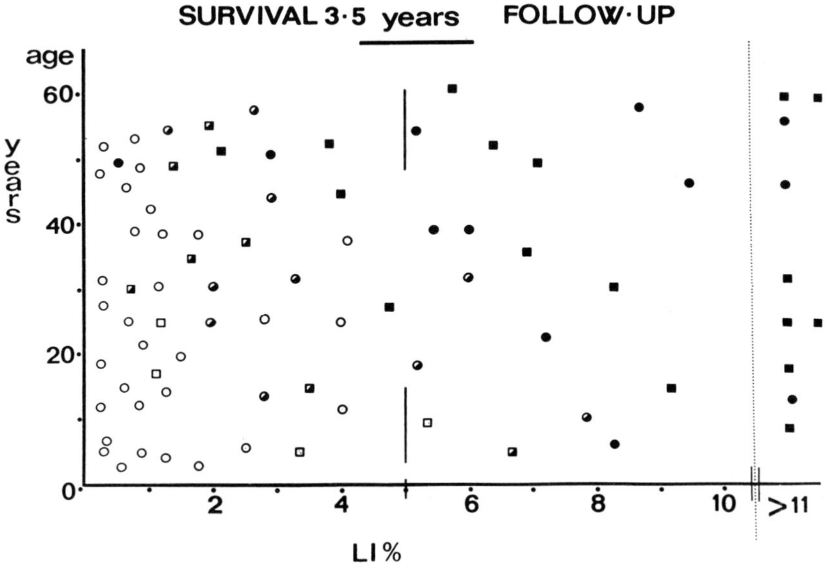

Fig. 1. Graphic representation of the relationships between age of patients at the time of stereotactic biopsy, LI values and outcome at 3-5 years follow-up. It has to be noted that the LI value detected in the fragments of tissue provided by serial stereotactic biopsy (2-7 samples performed along a stereotactic transtumoral trajectory) (4, 6). The circles represent mature astrocytoma. The squares represent anaplasic astrocytoma. The filled black symbols mean dead patients at 3-5 years follow-up. The half-filled symbols mean patients which underwent worsening of clinical conditions or widening of the tumor as detected by CT. The empty symbols mean patients whose clinical conditions and neuroradiological patterns may be considered stabilized. Patients whose LI resulted higher than 10% are grouped at the right of the dotted line, because their relatively uniform behaviour resembling that of glioblastoma (see text). The limit between patients which may expect relatively favourable outcome and patients with poor prognosis has been marked at 5% LI value.

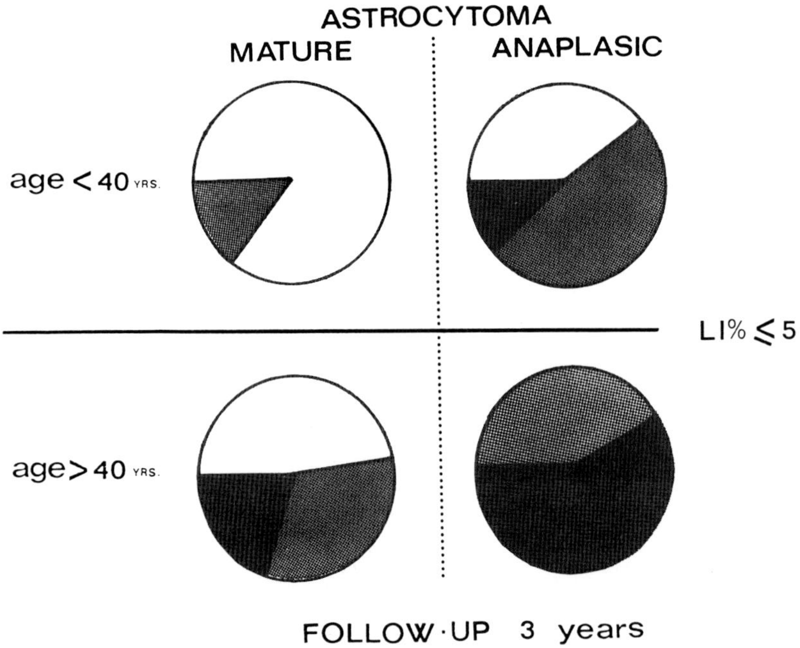

Fig. 2. Graphic representation of the outcome and survival at 3 years follow-up in patients affected by tumors whose LI values were below 5%. White space: stabilized patients; Gray space: patients which underwent clinical worsening of widening of the tumor at CT scan; black space: dead patients. Note the incidence of the age on the prognosis of patients belonging to the same histological class (mature astrocytoma on the left side and anaplasic astrocytoma on right side). Concerning the proliferative activity expressed by LI, all these tumors may be considered homogeneous and the different reactivity to the tumor aggressiveness may be depending by still unknown age realted antitumoral factors which modulate the behaviour either of mature than anaplasic astrocytoma.

ACKNOWLEDGEMENTS

This study has been supported in part by grant n. 87 01505.44 from Consiglio Nazionale delle Ricerche, Roma (Italy) and by the Associazione "P. Zorzi" for Neurosciences, Milano (Italy)

REFERENCES

1. Bookwalter JW, Selker RG, Schiffer L, Randall M, Iannuzzi D, Kristofik M (1986) Brain Tumor cell kinetics correlated with survival. J Neurosurg 65:795-798
2. Hoshino T, Wilson CB (1979) Cell kinetic analyses of human malignant brain tumors (gliomas) Cancer 44: 956-962
3. Piepmeier JM (1987) Observations on the current treatment of low-grade astrocytic tumors of the cerebral hemispheres. J Neurosurg 67: 177-181
4. Broggi G, Franzini A (1981) Value of serial stereotactic biopsies and impedance monitoring in the treatment of deep brain tumors. J Neurol Nerosurg Psychiatry 44:397-401
5. W.H.O. (1979) Histological classification of tumours of the nervous system. Geneve 1979
6. Franzini A, Broggi G, Allegranza A, Melcarne A, Ventura L, Costa A (1986) Cell kinetics of gliomas by serial stereotactic biopsy. Bas Appl Histochem 30: 203-207
7. Betti OO, Derechinski VE (1984) Hyperselective encephalic irradiations with linear accelerator. Acta Neurochir Suppl 33: 385-390
8. Kelly PJ, Bruce AK, Goerss BS, Earnest F (1986) Computer-assisted stereotactic laser resection of intra-axial brain neoplasms. J Neurosurg. 64: 427-439
9. Szikla G, Betti OO, Szenthe L, Schlienger M (1981) L'experience actuelle des irradiations stéréotaxiques dans le traitement des gliomas hémisphériques. Neurochirurgie 27: 295-298
10. Scanlon PW, Taylor WF (1979) Radiotherapy of intracranial astrocytomas: analysis of 417 cases treated from 1960 to 1969. Neurosurgery 5: 301-308

STEROID HORMONE RECEPTORS IN NEUROEPITHELIAL TUMORS: characteristics and biological role.

Butti G., Assietti R.,*Gibelli N.,** Scerrati M.,* Zibera C., Rolli M., Magrassi L.,**Roselli R., ***Sica G.,*Robustelli della Cuna G.

Department of Surgery - Neurosurgery, University of Pavia; *Division of Oncology, "Clinica del Lavoro" Foundation, Pavia; **Institute of Neurosurgery and ***Institute of Histology, "Cattolica" University of Rome, Italy.

INTRODUCTION.

The specific role of steroid hormone receptors has become of great relevance in the choice of the treatment modality of some tumors (1,10).

Recently, the presence of steroid hormone receptors has been demonstrated in intracranial tumors, expecially in meningiomas (7,9,12,21,23).

Neuroepithelial tumors (i.e. astrocytoma, anaplastic astrocytoma and glioblastoma) have been thought to express little content of steroid hormone receptors but data reported in literature are scanty and controversial (6,14,18). Till now, no data are available on the possible biological role of these receptors in neuroepithelial tumors.

We have studied the problem of the supposed influence of steroid hormones on the biological environment of neuroepithelial tumors.

Our results on the determination of steroid hormone receptors and on the influence induced by the specific hormones on the growth of cultured neuroepithelial tumors are reviewed in this paper.

STEROID HORMONE RECEPTORS DETERMINATION.

Matherials and Methods.

Tissue samples of neuroepithelial tumors were obtained at surgery from 57 unselected patients and immediately frozen in liquid nitrogen.

There were 25 glioblastomas, 18 anaplastic astrocytomas, and 14 astrocytomas.

The cytosol content of glucocorticoid (GR), androgen (AR), estrogen (ER) and progesterone (PR) receptors was determined on the frozen tissue specimens using the dextran-coated charcoal technique (DCC) as previously described (12). Results were expressed in femtomoles of receptor per milligram of cytosol protein (fmol/mg prot). The arbitrary cut-off value of more than 10 fmol/mg prot was considered to be indicative of a positive receptor value.

Results.

The results of the receptor content determination are shown in Table 1.

The range of Kd values of the different receptors are as follows: PR: 0.1-0.9 nM; AR: 0.1-0.9 nM; GR: 2-7 nM; ER: 0.5-0.7 nM.

Specific dexamethasone-binding proteins have been detected in 38.6% of the cases while AR are present in 21.6%. Only few tumor samples show ER or PR positivity.

Table 1
Mean concentration and positivity of steroid receptors

	N.	MEAN CONC. (range)	POSITIVITY %
GR	57	25.7 (10.5-76.1)	38.6
ER	57	17.7 (11.0-25.9)	8.8
PR	57	15.9 (14.7-17.1)	3.5
AR	51	24.1 (11.8-67.8)	21.6

Considering the influence of sex (Table 2), higher percentages of positivity are observed in females than in males.

Table 2.
Steroid receptors positivity (%) related to sex.

	MALE	FEMALE
GR	29.6%	46.7%
ER	3.7%	13.3%
PR	3.7%	3.3%
AR	4.1%	34.6%

It is remarkable that all patients but one presenting androgen-binding proteins belong to the female group. This group has been divided into four classes according to the age: <40, 40/50, 50/60, >60 years. 58.3% of the AR positive cases are older than 60 years, while GR and ER positivity is equally distributed in all classes.

The correlation between steroid receptor positivity and histological subtypes is shown in Table 3. It is interesting to note that the number of tumors that show GR positivity decreases in relation to the increase of histological malignancy; the opposite is noted for AR.

Table 3.
Steroid receptors positivity (%) related to histology.

	GLIOBLASTOMA	ANAPL. ASTRO.	ASTROCYTOMA
GR	20.0	33.3	50.0
ER	4.0	5.6	7.1
PR	0	5.6	0
AR	37.5	13.3	8.3

HORMONAL MODULATION ASSAY.

In our study reported above, the hormone binding proteins detected more frequently in neuroepithelial tumors are GR and AR. Based on this observation we studied the influence of specific hormones (dexamethasone for GR and testosterone for AR) on the growth of cultured neuroepithelial tumor cells.

Matherials and Methods.

Cell Culture. Fragments of tumor were obtained at surgery. Neoplastic specimens from 7 patients (2 anaplastic astrocytomas, case 1 and 2; 1 fibrillary astrocytoma, case 3; and 4 glioblastomas, cases 4-7) were minced, trypsinized, washed in saline and cultured into 25-sq cm tissue flasks with RPMI 1640 medium. Flasks were kept in a 5% CO_2 humidified atmosphere.

At confluence, cells were trypsinized and replated until an adequate number af them was available for cell growth experiments and steroid receptor determination.

Cell growth experiments. Cells were trypsinized and the suspension was passed through a fine sterile needle to obtain single viable cells.

They were seeded at an initial density of $5x10^3$ or $10x10^3$ cells/ml in 96-well flat-bottom tissue culture plates. 100 µl of cell suspension was added to each well and the plates were incubated at 37°C in a 5% CO_2 humidified atmosphere for 72 hours. Medium was then removed and replaced with 100 µl of medium in the presence or absence of different hormone concentrations. (Dexamethasone, Testosterone-acetate-Sigma).

Plates were fixed with methanol, stained with Giemsa 10% after four days, and each well was counted microscopically.

The mean number of cells per well (7 wells for control and 7 for each treatment) was then calculated and values were reported as percentage of controls. Student's t test was applied for statistical analysis.

Steroid hormone receptor determination on cell cultures. GR and AR were evaluated by a whole cell assay (16).

Cells were seeded at an initial density of $5x10^5$ cells/well in a 6-well tissue culture cluster and incubated in a 5% CO_2 humified atmosphere for three days.

The medium was then changed with a new one supplemented with 10% DCC-FCS (fetal calf serum with 0.25% (w/v) Norite activated charcoal for 16 hours followed by ultracentrifugation and passage through a 0.22 µm filter). Twenty-four hours later, cells were kept at 37°C for 15 min with RPMI 1640 without serum to remove most of the endogenous hormone and serum binding globulins.

Wells were incubated in duplicate, for 50 min, with 1 ml medium/dish containing the radiolabelled tracer (3H-dexamethasone, 3H-R1881 methyltrienolone for GR and AR determination respectively), in presence or absence of a 200-fold molar excess of unlabelled hormone. In the AR assay a 500 fold molar eccess of triamcinolone acetonide was added to avoid binding of 3H-R1881 to progesterone receptors without affecting androgen binding.

Medium was removed after incubation and wells were washed five times with cold physiological saline solution.

Radioactivity was extracted using 1ml of NaOH 1M and counted with the addition of 4 ml of scintillation fluid (Packard).

Results are expressed as fmol/µg DNA (3).

Results.

The results of the of AR and GR determinations were qualitatively similar both in tumor tissue and in cell cultures.

Six cases were tested for the activity on cell growth of scalar doses of dexamethasone (Table 4). Three of them were GR positive.

All the cultures showed significant growth inhibition after treatment with 50 µg/ml of dexamethasone.

Low drug concentrations (from 2 to 0.016 µg/ml) caused a significant enhancement of cell proliferation in two out of the three GR positive cultures. None of the GR negative cases was influenced by dexamethasone treatment at the same doses.

The influence of scalar doses of testosterone was tested on seven cell cultures as shown in Table 5. None of the seven cases tested showed specific androgen-binding proteins.

Table 4.
Dexamethasone activity on cell growth.

N.	CONCENTRATION (μg/ml)						GR
	50	10	2	0.4	0.08	0.016	
1	70±3*	107±12	124±3*	127±5*	130±10*	131±9*	POS
2	63±7*	133±14*	231±14*	296±4*	300±2*	156±3*	POS
3	22±5*	69±6*	88±10	94±9	94±5	109±11	POS
4	63±2*	94±4	99±6	101±15	96±5	97±3	NEG
5	50±5*	67±4*	73+9*	100±6	104±6	--	NEG
6	59±3*	85±9	102±6	105±8	100±3	102±3	NEG

*P<0.01

Table 5.
Testosterone activity on cell growth.

N.	CONCENTRATION (μg/ml)						AR
	50	10	2	0.4	0.08	0.016	
1	21±1*	72±6*	101±14	85±11	90±7	--	NEG
2	15±1*	42±1*	59±4*	97±6	93±6	--	NEG
3	--	28±3*	43±4*	86±1*	88±4	104±9	NEG
4	--	41±5*	63±3*	76±8*	90±11	92±7	NEG
5	--	16±1*	83±1*	100±4	97±2	96±1	NEG
6	--	4±1*	35±2*	56±4*	74±7*	101±6	NEG
7	--	53±5*	103±4	105±9	105±11	109±9	NEG

* P<0.01

Drug doses ranging from 50 to 10 μg/ml caused a significant cytotoxic effect in all the tumors tested. Growth inhibition was also present at the concentration of 2 μg/ml in six of them.

No significant drug action was noted at the concentration of 0.08 μg/ml except for case 6 in which inhibition was still present.

DISCUSSION.

In breast cancer, the presence of steroid receptors is correlated with the response to hormonal therapy in about 70% of cases (10), and both ablative and additive hormonal manipulations are considered valuable therapeutic strategies.

Studies on the determination of specific steroid hormone-binding proteins in brain tumors have been previously reported in the literature and meningioma has been the oncotype extensively examined. Nevertheless the results have shown frequent discrepancies in most series (4,13). Neuroepithelial tumors have been thought to be without hormone receptors (14) and only few report are published on this topic (5,6,14,18,19,24,25).

Hormonal treatment of human meningiomas has been suggested based on the detection of specific steroid-binding proteins and on the results obtained with "in vitro" functional assays. In studies reported in literature attention was focused on the presence of progesterone and estrogen receptors, and on the modification induced by the specific hormones (progesterone and estradiol) and antihormone drugs (RU-38486, tamoxifen, medroxyprogesterone acetate) on meningioma cell growth (8,11,15,17).

Only few reports (12,18,20,25) deal with the presence of androgen and glucocorticoid receptors and the functional significance of these receptors in brain tumors has not been tested yet. According to our studies, these two receptors are the most frequent steroid binding proteins observed in neuroepithelial tumors (21.6% positivity for AR and 38.6% for GR).

The number of patients studied in our series is one of the highest presented in the literature and, according to Blankestein (2), we think that the criteria adopted in the evaluation of receptor positivity are strict.

The correlation of this two receptors with the grade of anaplasia is noteworthy. Furthermore, it is noticeable that AR are present more frequently in females than in males. This difference appears relevant because epidemiological and experimental data suggested an androgen sensitivity of neuroepithelial tumors in some conditions (18).

After the demonstration of the presence of specific steroid hormone-binding proteins, we planned a study to verify if glucocorticoid and androgen hormones could have some effect on cell growth and if these effects could be related to the presence of specific receptors in cultures derived from these tumors.

In addition we studied early-passage cultured cells because cell lines may be not representative of the biological features of the original tumor.

It is noteworthy that the receptor status of tumor specimens is maintained in cell cultures.

The inhibition of cell growth obtained after treatment with high doses of both androgen and glucocorticoid hormones is probably not due to the receptor mediation, but seems to be related to other mechanisms (i.e. cell membrane alterations), as reported for other hormones (22).

Low doses of dexamethasone stimulate the growth in GR positive tumors so that the stimulation appears to be induced by the presence of the specific receptor.

The absence of drug activity in a GR positive case could be explained considering that, as in breast cancer, the receptor presence is a necessary but probably not sufficient condition to obtain a growth modulation.

The importance of the presence of the specific receptor for the growth modulation is enhanced by the absence of response in the GR negative cases.

This hypotesis is confirmed considering TA activity. In TA assay low drug concentrations are ineffective on all the AR negative cases.

These preliminary observations point out that dexamethasone and testosterone can influence the development of neuroepithelial tumors through the presence of the specific receptor and suggest that this type of study applied to malignant neuroepithelial tumors is usefull to obtain interesting informations on their biology and, probably, on new therapeutic strategies.

ACKNOWLEDGEMENTS.
This study is supported by the grant N° 87.01204.44 from the National Research Council, Rome, Italy.

REFERENCES.

1. Allegra JC, Lippman ME, Thompson EB, Simon R (1979) Relationship between the progesterone androgen and glucocorticoid receptors and response rate to endocrine therapy in metastatic breast cancer. Cancer Res 39:1973-1979.

2. Blankestein MA, Blaauw G, Lamberts SWJ, Mulder E (1983) Eur. J. Cancer Clin. Oncol., 19, 365.

3. Burton KA (1956) A study of the conditions and mechanism of the diphenylamine reaction for the colorimetric estimation of deoxyribonucleic acid. Biochem J. 62:315-323.

4. Cahill DW, Bashirelahi N, Solomon LW, Dalton T, Salcman M, Ducker TB (1984) Estrogen and progesterone receptors in meningiomas. J Neurosurg 60:985-993.

5. Concolino G, Giuffré R, Margiotta G, Liccardo G, Marocchi A (1984) Steroid receptors in CNS: estradiol (ER) and progesterone (PR) receptors in human spinal cord tumors. J Steroid Biochem 20:491-94.

6. Courriere P, Tremoulet M, Eche N, Armand JP (1985) Hormonal steroid receptors in intracranial tumors and their relevance in hormone therapy. Eur J Cancer Clin Oncol 21:711-714.

7. Glik RP, Molteni A, Fors EM (1983) Hormone binding in brain tumors. Neurosurg 13:513-519.

8. Grunberg SM, Daniels AM, Muensch H, Daniels JR, Bernstein L, Cortes V, Weiss MH (1987) Correlation of meningioma hormone receptor status with hormone sensitivity in a tumor stem-cell assay. J Neurosurg 66:405-408.

9. Hinton D, Mobbs EG, Sima AA, Hanna W (1983) Steroid receptors in meningiomas. A histochemical and biochemical study. Acta Neuropathol 62:134-140.

10. Horwitz KB, McGuire WL, Pearson OH, Segaloff A (1975) Predicting response to endocrine therapy in human breast cancer a hypothesis. Science 189:726-727.

11. Jay JR, MacLaughlin DT, Riley KR, Martuza RL (1985) Modulation of meningioma cell growth by sex steroid hormones in vitro. J Neurosurg 62:757-762.

12. Knerich R, Scerrati M, Butti G, Sica G, Zibera C, Silvani V, Robustelli della Cuna G, Rossi GF (1987) Steroid hormone receptors and intracranial tumors. In: Chatel M, Darcel F, eds. Brain Oncology. Dordrecht: Martinus Nijhoff, pp159-163.

13. Maiuri F, Montagnani S, Gallicchio B (1986) Estrogen and progesterone receptors in meningiomas. Surg Neurol 26: 435.

14. Markwalder TM, Zava DT, Markwalder RW (1983) Sexual steroid hormone receptors assay in human astrocytoma. Surg Neurol 20, 263.

15. Markwalder TM, Waelti E, Konig MP (1987) Endocrine manipulation of meningiomas with medroxyprogesterone acetate. Surg Neurol 28:3-9.

16. Natoli C, Sica G, Natoli V (1983) Two new estrogen supersensitiv of the MCF-7 human breast cancer cell line. Breast Cancer Res Treat 3:23-32.

17. Olson JJ, Beck W, Schlechte JA, Loh PM (1987) Effect of the antiprogesterone RU-38486 on meningioma implanted into nude mice. J Neurosurg 66:584-587.

18. Poisson M, Pertuiset BF, Hauw JJ, Philippon J, Buge A, Moguilewsky M, Philibert D (1983) Steroid hormone receptors in human meningiomas, gliomas, and brain metastases. J Neuro-Oncol 1:179-189.

19. Reganchary SS, Tiltzer LL, (1981) A study of dexamethasone receptor protein in human gliomas. J Surg Res 31:447-55.

20. Schnegg JF, Gomez F, Le Marchand-Berand T, de Tribolet N (1981) Presence of sex steroid hormone receptors in meningioma tissue. Surg Neurol 15:415-418

21. Schwartz MR, Randolph RL, Cech DA, Rose JE, Panko WB (1984) Steroid hormone binding macromolecules in meningiomas. Failure to meet criteria of specific receptor. Cancer 53:922-927.

22. Sica G, Natoli V, Marchetti P Piperno S, Iacobelli S (1984) Tamoxifeneinduced membrane alteration in human breast cancer cells. J Steroid Biochem 20:425-428.

23. Tilzer LL, Plapp FV, Evans JP, Stone D, Alward K (1982) Steroid receptor proteins in human meningiomas. Cancer 49:633-636.

24. Vaquero J, Marcos ML, Martinez R, Bravo G (1983) Estrogen- and progesterone-receptor proteins in intracranial tumors. Surg Neurol 19:11-13.

Cerebral gliomas. G. Broggi and M.A. Gerosa, editors

VASCULARIZATION AND ANGIOGENESIS IN BRAIN TUMORS

MARIA TERESA GIORDANA, MARIA CLAUDIA VIGLIANI
2nd Department of Neurology, University of Turin, Turin, Italy

The blood supply of a tumor is necessary for its progressive growth, for the delivery of chemotherapeutic drugs and imaging agents. In cerebral tumors the structural changes of the microvasculature are the anatomic basis of cerebral edema; ultrastructural studies of the blood vessels of human cerebral tumors have shown many pathologic features, including fenestration, widened interendothelial junctions, increased pinocytic vescicles, loosened and thickened and partly disrupted basement membranes (1, 2, 3, 4, 5, 6). These changes are thought to be related to alterations in the blood-brain barrier and to the increased vascular permeability.

On the contrary, in the brain immediately adjacent to the tumor mass (BAT) capillary permeability is reduced, as has been demonstrated by tracer studies (7); this may modify the delivery of chemotherapeutic agents to infiltrating tumor cells.

The ultrastructure of vessel wall basement membrane (BM) in malignant brain tumors has been studied, changes, such as irregularities of width, splitting in multiple layers, ruptures have been observed (1, 6). Even if it has not yet been ascertained whether anatomic changes of the BM are involved in the breakdown of the blood-brain barrier, thickening and tortuosity are believed to be more important than ruptures (1, 8) . At light microscopic level, the demonstration of BM has been obtained (9, 10, 11) by means of the immunostaining of laminin, a specific glycoprotein of basement membranes (12) . The laminin layer corresponding to the endothelial BM is thickened, multilayered and continuous around the vessels, even in poorly differentiated gliomas. The continuity of perivascular laminin in gliomas is thus not inconsistent with a broken blood-brain barrier.

Also the glial basement membrane is relatively intact, suggesting that the glial BM forms a relative barrier to the invasion of the perivascular space by glioma cells (11).

Because of the distribution of BM in neural tissue (13), the immunohistochemical demonstration of laminin gives information about the vascular pattern of intracerebral tumors: some oncotypes, such as oligodendrogliomas and ependymomas have a rich vasculature, others such as medulloblastomas and astrocytomas have a poor vasculature. In malignant gliomas the vasculature and the vessel walls show a number of abnormalities (5) which are well known to neuropathologists: increased number of large and small vessels, dilatations and narrowing of the lumina, varying degrees of endothelial proliferations, often resulting in the formation of convoluted glomerulus-like channels, formation of capillary teleangectasias, intimal proliferations in medium size vessels, thrombotic phenomena, fibrinoid necrosis in small arteries (5, 14). Vascular hyperplasia is considered a characteristic feature of malignant gliomas (15), and an expression of malignancy (6, 16). In cerebral gliomas of various type and grade of malignancy, vascular density and mean vessel diameter differ, as shown in a morphometric investigation (17). The role of vessel alterations in gliomas and their relation to tumor growth rate are not yet completely ascertained; the imbalance between the capacity to enter mitosis of endothelial cells and that of tumor cells (18) could be responsible for the development of necroses (5, 19).

Multiple interrelationships occur between a growing tumor and vessels of the host tissue. The hypothesis was formulated that solid tumors are angiogenesis-dependent. Tumors fail to grow beyond a few millimeters in diameter, unless new vessels, elicited from the host tissue, penetrate them (20, 21, 22). These conclusions are based on studies of tumor growth in the hamster cheek pouch and rabbit eye chamber. The same phenomenon has been observed in astrocytomas trasplanted in the rat brain (23): tumor growth can be divided in 3 stages: avascular, early vascular and late vascular, depending on the tumor diameter. When the mass reaches 1-4mm, signs of angiogenesis appear, as capillary buds and small immature capillaries; as the tumor enlarges, the central zone contains pleomorphic blood vessels exhibiting the ultrastructural abnormalities typical of glioma vasculature.

The isolation of angiogenic and mitogenic factors from tumors (24, 25) reveals the mechanisms through which tumors could induce the growth of new capillaries (26). One group of angiogenic factors (Heparin-binding endothelial growth factors) has been found mainly in neural tissue (27, 28); an angiogenic activity was demonstrated in cell cultures and extracts from meningiomas and glioblastomas (25, 29, 30, 31).

The relationship between endothelial proliferation, neovascularization and angiogenetic factors in gliomas remains unknown. On one hand, endothelial proliferation in glioblastomas is prominent; on the other hand neovascularization, as seen by light microscopy, usually follows rather than preceeds tumor infiltration (32, 33) inconsistently with previous general opinion (6, 34, 35). The vessel density increases when malignant glioma cells have completely invaded the cortex. Vascular glomeruli, which represent the most prominent expression of endothelial proliferation, occur in areas with a low vessel density; the computer-assisted three dimensional reconstruction of the vascular network shows that endothelial hyperplasia deforms the shape of the vascular tree of infiltrated cortex. The angiogenetic reaction to tumor invasion seems thus to be late, slow and incostant. In human malignant gliomas, endothelial hyperplasia is not synonym of neovascularization; it appears a reactive phenomenon to tumor growth (32, 33).

No significant rise in numerical density of microvessels around the tumor, and a drop in vascular density in tumor centre, were seen also in intracerebral tumors of rats produced by inoculation of neoplastic astrocytes (36).

Other contributions are needed to increase the understanding of the natural evolution of tumor vessels; such vessels may be an important target for new treatment modalities.

ACKNOWLEDGEMENTS

Supported in part by a grant of the Italian National Research Council (CNR), Special Project "Oncology", contract N° 87.01446.44, Rome, and by the Italian Association for Cancer Research (AIRC).

REFERENCES

1. Long DM (1970) J Neurosurg 32: 127-144
2. Moriyasu N, Nakamura S, Miyagami M, Morisawa S, Igusa N (1971) J Clin Elec Microsc 4: 771-784
3. Hirano A, Zimmerman HM (1972) Lab Invest 26: 465-468
4. Hirano A, Matsui T (1975) Hum Pathol 6: 611-621
5. Waggener D, Beggs JL (1976) Adv Neurol 15: 27-49
6. Weller RO, Foy M, Cox S (1977) Neuropath Appl Neurobiol 3: 307-322
7. Levin VA, Freeman-Dove M, Landahl HD (1975) Arch Neurol 32: 785-792.
8. Klatzo I (1979) In: Thomas Smith W, Cavanagh JB (eds) Recent Advances in Neuropathology. Churchill, Livingstone, pp 27-41
9. Giordana MT, Germano I, Giaccone G, Mauro A, Mighali A, Schiffer D (1985) Acta Neuropathol (Berl) 67: 51-57
10. Bellon G, Caulet T, Cam Y, Pluot M, Poulin G, Pytlinska M, Bernard MH (1985) Acta Neuropathol (Berl) 66: 245-252
11. McComb RD, Bigner DD (1985) J Neuropathol Exp Neurol 44: 242-253
12. Timpl R, Martin GR (1982) In: Furthmayr H (ed) Immunochemistry of extracellular matrix, vol 2: Applications. CRC Press, Boca Raton, Fla
13. Peters A, Palay SL, Webster H de F (1976) The fine structure of the nervous system. Saunders, Philadelphia
14. Manuelidis EE, Solitare GB (1971) In: Minkler J (ed) Pathology of the nervous system. McGraw-Hill, New York, vol 2, pp 2026-2071
15. McComb RD, Bigner DD (1984) Clin Neuropathol 3: 93-106
16. Russell D, Rubinstein LJ (1977) Pathology of tumours of the nervous system. 4th ed, Arnold, London
17. Seitz RJ, Wechsler W (1986) In: Walker MD, Thomas DGT (eds) Biology of Brain Tumour. Nijhoff, Boston, pp 131-137
18. Hoshino T, Townsend JJ, Murakoa I, Wilson CB (1980) Brain 103: 967-984

19. Tannock IF (1970) Cancer Res 30: 2470-2476

20. Warren BA (1968) Angiologica 5: 230-249

21. Folkman J (1974) Advan Cancer Res 19: 331-357

22. Gimbrone MA Jr, Leapman S, Cotran RS, Folkman J (1973) J Natl Cancer Inst 50: 219-228

23. Deane BR, Lantos PL (1981) J Neurol Sci 49: 67-77

24. Folkman J, Merler E, Abernathy G, Williams G (1971) J Exp Med 133: 275-288

25. Klagsbrun M, Knighton D, Folkman J (1976) Cancer Res 36: 110-114

26. Folkman J, Klagsbrun M (1987) Science 235: 442-447

27. Lobb RR, Fett JW (1984) Biochemistry 23: 6295-6298

28. Bohlen P, Baird A, Esch F, Ling N, Gospodarowicz D (1984) Proc Natl Acad Sci USA 81: 5364-5368

29. Kelly PJ, Suddith RL, Hutchison HT, Werrbach K, Haber B (1976) J Neurosurg 44: 342-346

30. Hirschberg H (1984) Neuropath Appl Neurobiol 10: 33-42

31. Lye RH, Elstow SF, Weiss JB (1986) In: Walker MD, Thomas DGT (eds), Biology of brain tumors. Nijhoff, Boston, pp 61-74

32. Schiffer D, Chiò A, Giordana MT, Mauro A, Migheli A, Vigliani MC. Congress of American Neuropathologists, Seattle, 1987

33. Schiffer D, Chiò A, Giordana MT, Mauro A, Migheli A, Soffietti R, Vigliani MC. Submitted to Acta Neuropathol

34. Nyström S (1960) Acta Pathol Microbiol Scand, Suppl 137, 49: 1-85

35. Murray KJ, White JG, Douglas SD (1980) Surg Neurol 14: 53-58

36. Luthert PJ, Lantos PL (1985) Neuropath Appl Neurobiol 11: 461-473

Cerebral gliomas. G. Broggi and M.A. Gerosa, editors

IMPACT OF TUMOR PHYSIOLOGY AND ITS MODIFICATION ON BRAIN TUMOR THERAPY

PETER C. WARNKE, DENNIS GROOTHUIS*

Abteilung für Stereotaktische Neurochirurgie, Neurochirurgische Universitätsklinik, D - 6650 Homburg/Saar
* Div. of Neurology, The Evanston Hospital, Northwestern University, Evanston, Ill. 60201, USA

INTRODUCTION

The efficacy of brain tumor therapy employing cytotoxic drugs and/or monoclonal antibodies highly depends on the physiological characteristics of a given tumor i.e. its regional blood flow, its capillary permeability and its total capillary surface as well as the size of the extracellular distribution space for a specific compound (8). After numerous empirical chemotherapy trials have been conducted with only little effect on the course of the disease and with even less understanding of the underlying reasons which produced responses in some cases and none in others, several investigators focussed on the pharmacokinetic parameters influencing drug delivery to brain tumors (11,6). At this point in time methods became available to quantitatively measure two of the above parameters in experimental brain tumors: Regional cerebral blood flow (F) and capillary permeability (K) (1). Using Quantitative Autoradiography (QAR) or Positron Emission Tomography (PET) it became possible to measure rCBF and capillary transfer rates for different marker compounds in experimental tumors as well as in patients (7,9). These investigations strongly indicated that drug delivery to brain tumors was restricted by the partially intact blood-brain barrier of tumors and the reduced regional blood flow. Whereas the first limitation of drug delivery is mainly due to water-soluble drugs, the latter mainly influences delivery of lipophilic drugs. As a consequence studies were inaugurated to develop means to increase drug delivery to brain tumors(13,12). This paper will focus on the impact of regional tumor blood flow and blood-to-tissue transport in tumor models on future chemotherapy concepts and we will review our approaches to experimentally increase drug delivery to brain tumors by alteration of blood flow using adenosine or alteration of capillary permeability employing hyperosmotic blood-brain barrier disruption (HBBBD).

MATERIAL AND METHODS

Tumor models studied were the TE-671 cell line, the D-54MG human glioma cell line and transplacentally induced oligodendrogliomas and mixed astrocytomas after injection of ethylnitrosourea (ENU) into pregnant female Sprague-Dawley rats. Both, TE-671 and D-54MG were studied after intracerebral injection of a tumor cell suspension into nude rats.

Experimental Procedures

Combined F and K Measurements (TE-671) Regional blood flow and blood-to-tissue transport were measured usind double-label quantitative autoradiography and 131I-iodoantipyrine (IAP) and 14C-alpha-aminoisobutyric acid (AIB) as respective markers in nude rats bearing intracerebral xenotransplants of TE-671 medulloblastomas. Technical details concerning tissue preparation, radioactive standards and autoradiographs have been desribed previously (4). rCBF was calculated employing Kety-Schmidt equations, implying a λ-value of 0,8, as described by Sakurada (17). The blood-to-brain transfer rate (K) for AIB was calculated according to Patlak, et al (2). The net extraction fraction (En) was calculated using the expression En = K/FV(f) V (f) represents the fractional amount of blood from which AIB could be extracted and in this set of experiments could be assumed to be 0,5.

Alteration of blood-brain barrier (ENU-induced Gliomas) Rats with transplacentally induced gliomas were used for unilateral hyperosmotic blood-brain barrier disruption (HBBBD) when 7 months old. HBBBD was performed using an intracarotid infusion of 1.6 osmolar mannitol warmed to body temperature. Mannitol was infused retrogradely into the external carotid artery as described by Rapoport (15). The infusion rate was .12 ml/sec for 30 seconds. Thirty seconds after HBBBD 14 C-AIB was injected via a femoral venous catheter and allowed to circulate for ten minutes. Then the animal was decapitated and the brain further processed for QAR as described previously (4). A unidirectional blood-to-brain transfer rate for AIB was calculated according to Patlak (2) after correction for vascular space.

Alteration of regional tumor blood flow (D-54MG gliomas)

Symtomatic nude rats bearing intracerebral xenotransplants of D-54MG human gliomas were randomly allocated to either the treatment or control group. The treatment group received an i.v.-infusion of adenosine at a dose rate of 10 mol/kg/min and the control group received 0.9 % saline at the same infusion rate. After 4 minutes of infusion 14 C- IAP was infused with an increasing ramp schedule for 30 seconds as described previously (1). After the IAP-infusion the animals were decapitated and the brains were processed for QAR. F was calculated as described by Sakurada, et al (17) with a λ of 0.8. With the exception of the BBB-disruption experiments, which were done under anesthesia (halothane, nitrous oxide, oxygen: 1.5:70:30v/v/v) all animals were awake and the physiological status (blood pressure, blood gases and ph) was monitored and kept in physiological range.

RESULTS

Combined Blood Flow and Capillary Permeability Morphologically there were two distinctly different patterns of growth in TE-671 intracerebral xenotransplants. The vast majority of the twenty tumors appeared to be solid intraparenchymal masses, whereas three tumors exposed a perivascular growth pattern. The results are summarized in Table 1 showing whole tumor values for F and K for each tumor and the resulting net extraction fractions. The averaged mean transfer constant (K) of AIB was 6.8 $\pm$ 0.9 ml/100g/min which was significantly higher than values from brain adjacent to tumor (0.7 $\pm$ 0.5 ml/100g/min) or more dictant tumor-free brain ($p < 0.001$, ANOVA). Nevertheless there were three tumors (Nos 6, 9 and 12) which showed conspicuously lower K-values (mean 0.4 $\pm$ 0.06 ml/100g/min)and microscopically represented "perivascular" tumor growth. Interestingly the mean K-value of the "perivascular" tumors was not significantly different from that of tumor-free brain ($p > 0.05$, Student's t-test). Table 1 may also serve to demonstrate the variation of capillary permeability, expressed as K, between different tumors as well

Table 1

Blood flow and capillary permeability values for all TE-671 tumors and corresponding extraction fraction. Values are given as means ± standard deviation for individual tumors and as mean ± standard error for all tumors.

Tumor-No.	rCBF ml/100g/min	Capillary Transfer ml/100g/min	E_N
1	90.5 ± 44.8	8.3 ± 4.9	0.18
2	70.4 ± 20.2	7.7 ± 3.5	0.22
3	48.5 ± 12.9	7.0 ± 3.4	0.29
4	26.9 ± 9.5	7.1 ± 3.9	0.53
5	38.1 ± 14.8	6.7 ± 2.6	0.35
6	84.0 ± 21.6	0.5 ± 1.2	0.01
7	64.1 ± 22.0	4.8 ± 5.1	0.15
8	34.2 ± 24.1	7.5 ± 3.7	0.44
9	66.8 ± 11.0	0.4 ± 0.5	0.01
10	37.0 ± 20.0	17.8 ± 13.1	0.96
11	25.4 ± 24.9	9.7 ± 7.4	0.77
12	100.7 ± 23.6	0.3 ± 0.7	0.006
13	28.5 ± 16.9	8.3 ± 5.9	0.58
14	40.1 ± 16.7	10.9 ± 4.3	0.54
15	34.4 ± 36.0	6.9 ± 3.5	0.40
16	30.8 ± 23.5	6.0 ± 2.9	0.38
17	12.3 ± 9.5	7.1 ± 3.9	1.0
18	4.7 ± 5.5	6.7 ± 3.1	1.0
19	32.1 ± 18.6	7.6 ± 3.6	0.47
20	7.9 ± 8.9	7.1 ± 4.3	1.0
All tumors	43.9 ± 6.1	6.8 ± 0.9	0.46 ± 0.34

as within one and the same tumor as can be seen from the standard deviation. Regarding only the "solid" tumors there was an almost four-fold difference in capillary permeability between individual tumors and a 12-45-fold higher mean K-value as compared to normal brain. The values for whole tumor blood flow (F) are also shown in Table 1. The mean rCBF of all tumors was 43.9± 6.1 ml/100g/ min and not significantly different from blood flow in ipsilateral corpus callosum (41.6 ± 4.5 mi/100g/min). $P > 0.05$, paired t-test. Nevertheless F in "solid" tumors (36.8± 5.3 ml/100g/min)was significantly lower than F in "perivascular" tumors(83.8 ± 13.8 ml/100g/min)($p < 0.05$, ANOVA) or in contralateral corpus callosum (52.9 ± 5.5 ml/100g/min). "Perivascular" tumors on the other hand showed blood flow values higher than tumor free white matter but still lower than normal cortex ($p < 0.05$, ANOVA). Simultaneous measurement of blood flow and capillary transport rates for a given compound also allowed to calculate the net extraction fraction (En) of the tumor for this particular compound if certain requirements are met (4). Thus the net extraction fraction of AIB could be calculated for each individual tumor as shown in Table 1. En represents the relative portion of blood-borne AIB which is removed from blood during multiple capillary passages thus reaches the tumor extracellular space. Again En was different for the two morphological tumor entities: Whereas the "perivascular" tumors showed a mean En of 0.009 ± 0.001, the "solid" tumors exhibited a completely different physiology with a mean En of 0.54 ± 0.07.

Alteration of Blood-Brain Barrier

Histological examination of the 28 ENU-induced tumors that were studied revealed 22 oligodendrogliomas and 6 mixed gliomas. The morphological appearance of the tumors resembled those found in other studies (10). The capillary transfer constant K for AIB was measured in different areas of tumor-free brain and the tumor itself. As all animals were bearing multiple tumors and tumors were located in the non-disrupted hemisphere also, those served as control group for the tumors that had undergone HBBBD. The results are summarized in Table 2. HBBBD was confined to the middle cerebral and anterior cerebral artery territory of the hemisphere ipsilateral to the

Table 2

K-values of ENU-tumors in the disrupted and non-disrupted hemispheres. Values are given as means ± standard deviation for individual tumors and as mean ± standard error for all tumors. Dimensions for K are ml/g/min x 10^3.

Nondisrupted Hemisphere		Disrupted Hemisphere	
Tumor No.	Whole Tumor	Tumor No.	Whole Tumor
1	4.9 ± 3.7	1	13.9 ± 2.8
2	1.5 ± 1.9	2	3.9 ± 3.3
3	5.1 ± 2.4	3	40.2 ± 8.6
4	1.9 ± 1.6	4	23.2 ± 6.8
5	3.3 ± 2.4	5	11.4 ± 6.2
6	4.9 ± 2.4	6	8.3 ± 4.5
7	2.0 ± 2.2	7	15.4 ± 3.8
8	20.4 ± 8.8	8	3.9 ± 2.5
9	1.1 ± 2.8	9	6.9 ± 5.3
10	7.8 ± 2.7	10	0.1 ± 1.1
11	1.3 ± 1.8	11	16.1 ± 3.6
12	2.2 ± 2.3	12	11.5 ± 3.4
13	22.4 ± 8.9	13	17.4 ± 10.7
14	14.5 ± 4.0		
15	13.4 ± 7.5		
All tumors	7.1 ± 1.8	All tumors	13.3 ± 2.9

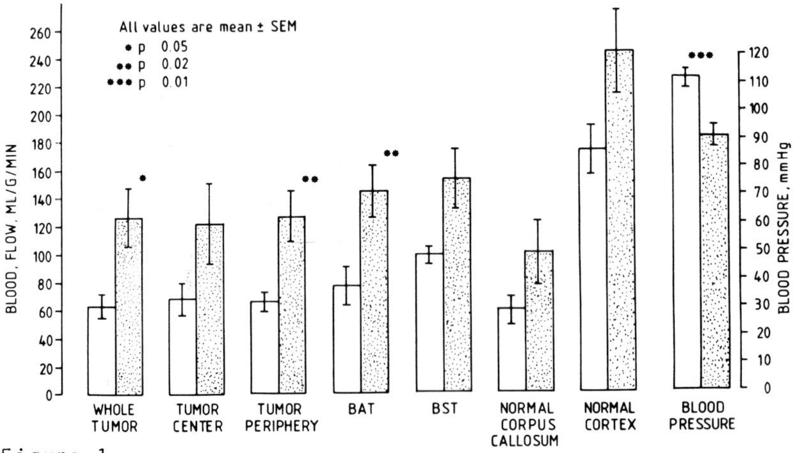

Figure 1:
Bar diagram of F-values in adenosine treated (dotted bars) and control animals (open bars) bearing D-54MG gliomas. The two right bars represent mean blood pressure values. Significant increases in F on the statistical level indicated were obtained only in whole tumor, tumor periphery and brain adjacent to tumor (BAT), not in brain surrounding tumor (BST) or normal brain.

hyperosmolar infusion. The variability of permeability changes in normal brain induced by HBBBD was conspicuous as can be seen from the range of K from 0.004 to 0.112 ml/g/min in normal cortex with a mean value of 0.050 ± 0,011 ml/g/min SE. The variability was not only pertinent to gray matter, but could be seen in ipsilateral white matter as well. The range of K-values in ipsilateral corpus callosum was from 0.001 - 0.057 ml/g/min with a mean value of 0.017 ± 0.005 ml/g/min. The K-values for individual tumors in the disrupted and non-disrupted hemisphere are shown in Table 2. There was a significant 35-fold increase in capillary permeability of cortex in the disrupted hemisphere over the non-disrupted and a 20-fold increase in disrupted corpus callosum versus non-disrupted. Comparing the capillary transfer constants of tumors in the disrupted hemisphere (mean K-value 0.013 ± 0.003 ml/g/min) with those in the non-disrupted hemisphere (mean K-value .007 ± .002 ml/g/min) did not show significant differences ($P > 0.01$, ANOVA) due to the variation in individual tumor values.

Furthermore the K-value of disrupted cortex in the ipsilateral hemisphere was significantly higher (mean K-value 0.05 ± 0.02 ml/g/min) than the mean tumor K-value of 0.013 ± 0.003 ml/g/min/ $p < 0.01$, Student's t-test).

Alteration of Regional Tumor Blood Flow The D-54MG gliomas histologically presented as anaplastic astrocytomas. There were five tumors in the control group and nine tumors in the adenosine treatment group. Adenosine infusions with an increasing dose regimen led to a decrease of systemic blood pressure (BP) in normal Fischer-344 rats. In the nude rats bearing D-54 MG-gliomas the control animals had a mean BP of 112 ± 6 mmHg as compared to 95 ± 9.4 mmHg in the adenosine treated animals. The difference was significant on the $p < 0.001$ level (Student's t-test). Blood flow measurements are summarized in Fig. 1 showing regional F-values for various tumor regions as well as tumor-free brain. For whole tumor blood flow there was a significant increase of F in the treated animals with a mean F of 117.6 ± 20.8 ml/g/min. Further regional analysis revealed that the differences in F between the two groups were not significant in tumor center, but in

tumor periphery and in brain adjacent to tumor (BAT) with a mean F of 143.1 $\pm$ 18.9 ml/100g/min in the adenosine group and 82.9 $\pm$ 11.9 ml/100g/min in the control group ($p < 0.02$, Student's t-test).
When tumor-free cortex and corpus callosum of both groups was compared there was no significant difference in F-values ($p > 0.05$, ANOVA) indicating a selective increase of F by i.v.-adenosine in the tumors only.

DISCUSSION

These studies of tumor physiology and drug delivery parameters in experimental brain tumors have been done to evaluate the impact of tumor physiology on brain tumor chemotherapy. As drug delivery to brain tumors depends on the integrity of the blood-brain barrier as well as on blood supply of the tumor it has become obvious that studies of tumor physiology may provide a rational basis for chemotherapy approaches (5). In order to assess the impact of the above mentioned parameters we simultaneously studied blood flow and capillary permeability in the TE-671 medulloblastoma tumor modell (19). Compared to all brain tumor models studied previously TE-671 had the highest K-values so far, whereas regional blood flow was comparatively low. Still this constellation resulted in high extraction fractions (En). Nevertheless there were three morphologically "perivascular" tumors showing extremely low K-values approaching those of normal brain and corresponding high blood flow also in the range of normal brain. Strikingly the TE-671 model had drug delivery parameters (increased capillary permeability by more than a log unit, low blood flow) which might explain why medulloblastomas as an entity respond to treatment with water-soluble drugs which other primary brain tumors rarely do. The validity of the TE-671 model for therapy studies has recently been evaluated by measurements of capillary permeability in patients bearing medulloblastomas. Interestingly the increase of capillary permeability in these patients was in the same range as in TE-671, making it a valuable model for further therapy studies. Also it can be concluded that individual tumors as for example the "perivascular" ones may exhibit conspicuously different physiological characteristics calling for a different therapeutic approach.

As in most tumor models as well as in patient studies drug delivery esp. of water-soluble drugs is impaired by the partially intact BBB we studied HBBBD as a means to increase capillary permeability in ENU-induced gliomas. The model was chosen because of its naturally low K-values (3) for studies of HBBBD in other tumor models showing increased capillary permeability already before disruption failed to demonstrate significant effects on tumor permeability (12,18). Nevertheless capillary permeability of normal brain could be increased significantly in those studies. Therefore the ENU-tumors seemed to be the ideal model. Although the mean K-value of ENU-tumors in the disrupted hemisphere was almost twice that of tumors in the non-disrupted hemisphere, there was no significant increase of capillary permeability overall due to the variability of tumor K-values. In normal cortex and corpus callosum a 20-35-fold increase of K could be seen resulting sometimes in higher K-values than in the tumor. Thus HBBBD in some instances might turn out to be counterproductive by increasing delivery of cytotoxic agents to normal brain without increase of delivery in the tumor. As this has been found by other authors, too and in the light of recent studies showing cerebral infarctions as a side effect of the method (16), it should only be used in a careful, randomized and controlled fashion. To increase delivery of lipophilic drugs to brain tumors an increase of tumor blood flow seems to be suitable. We used adenosine in D-54MG gliomas because of previous reports in the literature (14). Adenosine is water- soluble so its passage across the BBB is restricted unless BBB-integrity as in D-54MG is destroyed. Using i.v. adenosine we were able to selectively increase F in the tumor and in brain adjacent to the tumor by almost 100 %. Although the potent vasodilator adenosine produced a significant decrease in blood pressure, the values staid within the physiological range. The impact of this study on delivery of lipophilic drugs (e.g. nitrosoureas) is enormous as agents with a high octanol/water partition coefficient easily cross the BBB and access to the tumor depends on regional blood flow.

So if the systemic side effects of adenosine could be tolerated this seems to be a promising tool to selectively manipulate brain tumor physiology.

The studies conducted elucidated the impact of brain tumor physiology on drug delivery in several ways:

1. Delivery of chemotherapeutic agents depends on regional capillary permeability and blood flow of a given tumor which is different in tumors of different histology, but also varies between individual tumors of the same histology as can be seen from our studies in medulloblastomas, which turned out to be a very permeable tumor model thus favoring further chemotherapeutic approaches.

2. Manipulation of tumor physiology which is intended to increase drug delivery raises complex issues and should be tested carefully in tumor models before used in patients. Though we were not able to increase BBB-permeability in tumors by means of HBBBD, we successfully disrupted the normal brain microvasculature causing possible damage to healthy brain.

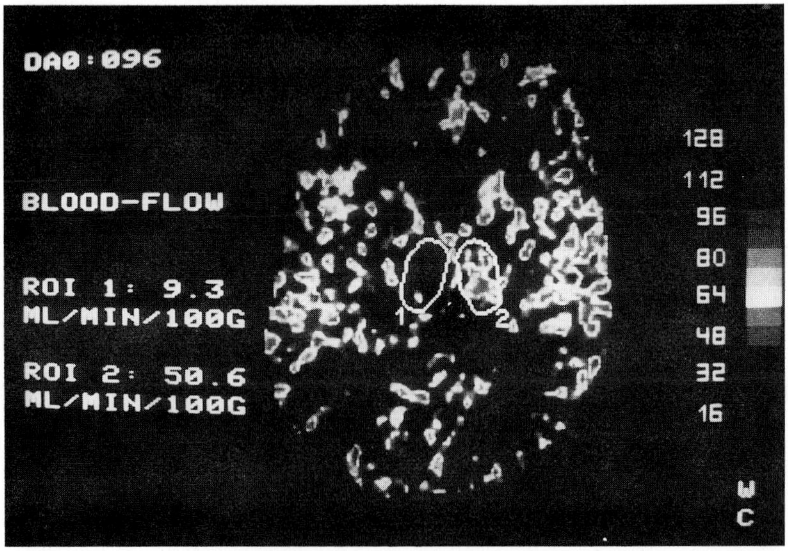

Figure 2:
RCBF-image obtained with the stable xenon CT-method in a patient bearing a left thalamic astrocytoma grade IV. The blood flow values are given in a gray scale on the right. Note decreased rCBF in the tumor (9.3 ml/100g/min) as compared to the contralateral thalamus (encircled region, 50.6 ml/100g/min).

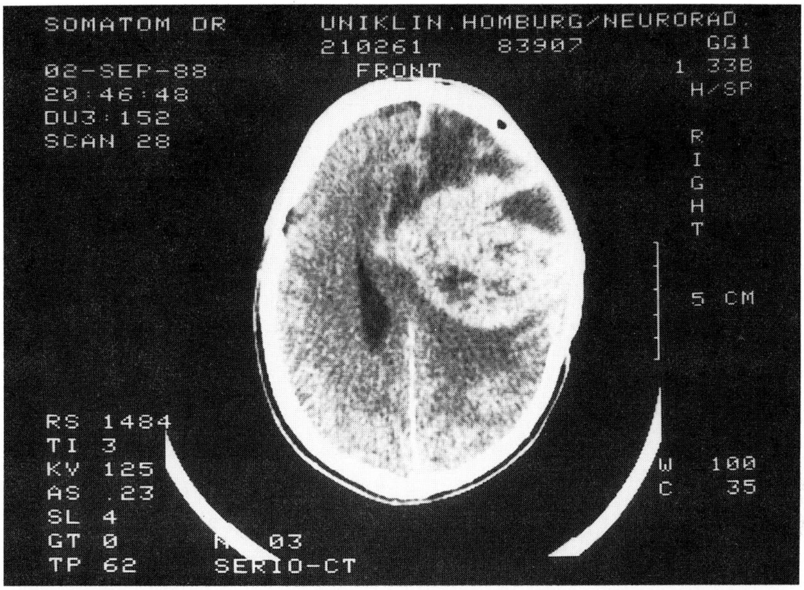

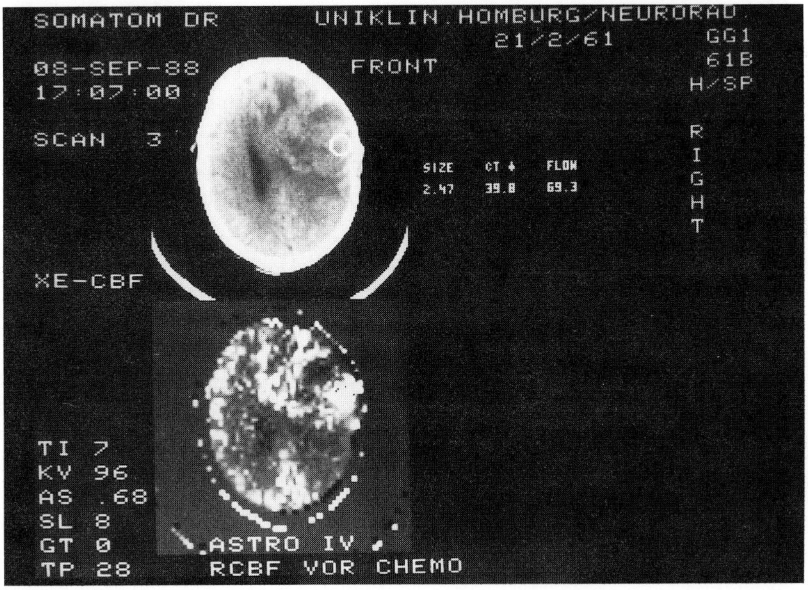

Figure 3:
Contrast-enhanced CT-image and rCBF-image of a patient with a recurrent astrocytoma grade IV. Note from the rCBF-image that tumor blood flow in the encircled region of interest was 69.3 ml/100g/min and higher than that of surrounding brain as can be seen from the gray scale encoded image.

Selective manipulation of blood flow in tumors was possible and may turn out therapeutically efficient for lipophilie drugs, but still it needs to be documented that adenosine crosses the BBB in less permeable tumor models which reflect the physiological range of human tumors.
The studies of drug delivery parameters in experimental brain tumors can be extrapolated to human tumors since the advent of new imaging techniques. Using elegang PET-or CT-techniques in vivo measurements of blood flow and capillary permeability became possible (Fig. 2 and 3). As preliminary studies of the above mentioned drug delivery parameters in human gliomas also showed a tremendous variability, only an individualized chemotherapy protocol seems to be appropriate for each single patient and careful planning of therapy modalities including modification of tumor physiology should be based on quantitative assessment of the physiologic parameters of a tumor.

REFERENCES

1. Blasberg RG, Groothuis DR, Molnar P: The application of quantitative autoradiographic measurements in experimental brain tumors. Semin Neurol 1:203-221, 1981

2. Blasberg RG, Patlak CS, Fentermacher JD: Transport of alpha-aminoisobutyric acid across brain capillary and cellular membranes, J Cereb Blood Flow Metab 3: 8-32, 1983

3 Blasberg RG, Kobayashi T, Horowitz M: Regional blood-to-tissue transport in ethylnitrosourea-induced brain tumors. Ann Neurol 14: 202-215, 1983

4. Blasberg RG, Molnar P, Groothuis DR, Patlak CS, Owens E, Fenstermacher JD: Concurrent measurements of blood flow and transcapillary transport in ASV-induced experimental brain tumors: implications for brain tumor chemotherapy. J Pharmacol Exp Ther 231: 724-735, 1984

5. Blasberg RG, Groothuis DR: Chemotherapy of brain tumors: physiological and pharmacokinetic considerations. Semin Oncol 13: 70-82, 1986

6. Fenstermacher JD, Blasberg RG, Patlak CS: Methods for quantifying the transport of drugs across brain barrier systems. Pharmacol Ther 14:217-248, 1981

7. Groothuis DR, Molnar P, Blasberg RG: Regional blood flow and blood-to-tissue transport in five brain tumor models. In: Rosenblum M, Wilson C (eds): Progress in Experimental Tumor Research, S. Karger, Basel, Vol 27:132-152, 1984

8. Groothuis DR, Blasberg RG: Rational brain tumor chemotherapy. Neurologic Clinics 3(4):801-816, 1985

9. Hawkins RA, Phelps ME, Huang SC: A kinetic evaluation of blood-brain barrier permeability in human brain tumors with (68 Ga) EDTA and positron emission tomography. J Cereb Blood Flow Metab 4:507-515, 1984

10. Koestner A, Swenberg JA, Wechsler W: Transplacental production with ethylnitrosourea of neoplasms of the nervous system in Sprague-Dawley rats. Am J Pathol 63:37-50, 1971

11. Levin VA, Landahl HD, Freeman-Dove MA: The application of brain capillary permeability coefficient measurements to pathological conditions and the selection of agents which cross the blood-brain-barrier. J Pharmacokinet Biopharm 4:499-519, 1976

12. Nakagawa H, Groothuis DR, Blasberg RG: The effect of graded hypertonic intracarotid infusions on drug delivery to experimental RG-2 gliomas. Neurology 34:1571-1581, 1984

13. Neuwelt EA, Rapoport SI: Modification of the blood brain barrier in the chemotherapy of malignant brain tumors. Fed Proc 43:214-219, 1984

14. Panther LA, Baumbach GL, Bigner DD, Piegors D, Groothuis DR, Heistad DD: Vasoactive drugs produce selective changes in flow to experimental brain tumors. Ann Neurol 18:712-715, 1985

15. Rapoport SI, Hori M, Klatzo F: Testing of a hypothesis for osmotic opening of the blood-brain barrier. Am J Physiol 223:323-331, 1972

16. Suzuki M, Iwasoki Y, Yamomoto T, Kanno H, Kudo H: Sequelae of osmotic blood-brain barrier opening in rats. J Neurosurg 69:421-428, 1988.

17. Sakurada O, Kennedy C, Jehle J: Measurement of local cerebral blood flow with iodo (14C) antipyrine. Am J Physiol 234:59-66, 1978

18. Warnke PC, Kuruvilla A, Mikhael M, Groothuis DR: The effect of hyperosmotic blood-brain barrier disruption on capillary permeability in normal and tumor-bearing dogs. Neurology 35 (suppl 1): 289, 1985

19. Warnke PC, Friedman HS, Bigner DD, Groothuis DR: Simultaneous measurements of blood flow and blood-to-tissue transport in xenotransplanted medulloblastomas. Cancer Research 47:1687-1690, 1987.

IMMUNOLOGICAL ASPECTS OF GLIOMAS

STEFAN BODMER, CHRISTINE SIEPL AND ADRIANO FONTANA
Section of Clinical Immunology, University Hospital, Häldeliweg 4, CH-8044 Zürich (Switzerland)

INTRODUCTION

The view of the brain to be an immunologically privileged site in which immune surveillance is only poorly possible is mainly based on the early demonstration that xenografts of a mouse sarcoma implanted into rat brain continued to grow, whereas the same tumor was rejected when transplanted subcutaneously (1,2). However, intact effector functions of the immune system within the brain has subsequently been shown by the rejection of allogenic skin grafts implanted into the brain of rabbits that were presensitized with an orthotopic skin allograft (3). Moreover, the demonstration of lymphocyte infiltrates in the brain tissue of patients with viral and autoimmune encephalitis (4) as well as the successful transfer of experimental autoimmune encephalitis by myelin basic protein specific T cells (5) indicated that at least the effector cells of the immune system may overcome the anatomical barrier (blood brain barrier) and act efficiently within the central nervous system (for review see references 6,7).

The present article will give a short summary on the immunological status of glioblastoma patients and take a closer look at results of in vitro studies investigating immunoregulatory factors secreted by cultured human glioblastoma cells. Of particular interest will be the biochemical and biological characterization of an immunosuppressive factor, termed glioblastoma derived T cell suppressor factor (G-TsF), which interferes with T cell growth. This factor is a member of the transforming growth factor-beta (TGF-beta) gene family and might have an important intrinsic function in tumor cell growth and prevention from immune-mediated anti-tumor responses.

IMMUNE STATUS OF GLIOBLASTOMA PATIENTS

Patients with glioblastoma multiforme, the most malignant form of the gliomas, normally die within one year despite surgery, radiotherapy, and chemotherapy. Impaired cellular immunity in these patients has been well documented by a number of investigations (for review see reference 7). The depressed cell-mediated immunity is evidenced by cutaneous anergy, diminished number of T cells, and markedly reduced blastogenic responsiveness of peripheral blood lymphocytes exposed in vitro to mitogens or in mixed lymphocyte cultures. Uncharacterized factors which inhibit the mitogen- and antigen-induced proliferation of normal lymphocytes have been observed in the patient serum before but not after tumor removal (8) as well as in the cyst fluid of glioblastoma (9). This led to the suggestion that glioblastoma cells may release immunosuppressive factors. Moreover, the in vitro activation of tumor infiltrating T cells which were eluted from glioblastoma tissue has been found to be impaired, a finding which may be explained by the pre-exposure of the cells to immunosuppressive factors.

A positive correlation between cellular immune suppression and degree of tumor anaplasia has been reported (10). However, the degree of cellular immune suppression at the time of diagnosis had no prognostic value for the patient's outcome. Glioblastoma patients generally have normal humoral parameters and serum immunoglobulin levels. Although in most glioblastoma patients humoral immune responses to their tumors develop (11), there is little evidence of significant cellular anti-tumor immune responses. Impaired tumor surveillance may also originate from lymphokine-induced production of mucopolysaccharide coats by glioblastoma cells that non-specifically protects from cellular anti-tumor effector mechanisms (12).

CHARACTERIZATION OF GLIOBLASTOMA CELLS AND PRODUCTION OF IMMUNOSUPPRESSIVE FACTORS

Histologically, malignant gliomas are very heterogeneous tumors and this heterogeneity has been confirmed by phenotypic and genotypic studies of cultured glioma cell lines. Cultured human glioblastoma cells are characterized by the presence of microfilaments as well as by the biochemical markers glial fibrillary acidic protein (GFAP) and S-100 protein. Most commonly the cells are negative for galactocerebroside (GC) while being positive for fibronectin (FN). Considerable efforts were undertaken to identify one or several tumor-specific or tumor-associated antigens which would allow to generate tumor-specific antibodies. Comparable to tumor cells in situ also cultured cells may express neuroectodermal antigens shared with melanomas and neuroblastomas. Furthermore, most glioblastoma cell lines carry DR-like antigens of the Major Histocompatibility Complex (13). Cultured glioblastoma cells either constitutively release or can be induced to secrete various immunoregulatory factors. Glioblastoma cell lines have been found to secrete factors with interleukin-1 (IL-1)-like activity (14,15), as well as factors with interleukin-3 (IL-3)-like activity (16). Furthermore, cultured glioblastoma cells release interferons (17) and prostaglandins (18). Recent studies showed that human glioblastoma cells (line 308) also secrete the B cell stimulatory factor-2 (BSF-2/IL-6, unpublished results) and that the BSF-2/IL-6 mRNA can be induced by IL-1beta in glioblastoma and astrocytoma cell lines (19).

As reported in 1984 by this laboratory (15), another immunoregulatory mediator was also detected in the supernatant (SN) of human glioblastoma cells. This factor was initially characterized by the property to inhibit the lectin induced proliferative response of mouse thymocytes and to block the IL-2 or antigen induced growth of cloned mouse T cell lines. The factor, termed glioblastoma derived T cell suppressor factor (G-TsF), was also demonstrated to inhibit the generation of cytotoxic T cells (CTL) in a mixed lymphocyte culture (MLR). Purification and cloning of G-TsF revealed its relationship to the transforming growth factor beta gene family (20,21), structurally identical to TGF-beta-2 which was independently purified from a prostatic adenocarcinoma cell line (31).

The release of immunosuppressive factors into the conditioned medium is not a general property of cultured tumor cells. The SN of three neuroblastoma, two melanoma, and one rhabdomyosarcoma cell line failed to suppress the ConA-response of thymocytes or had only weak inhibitory effects (15). In the presence of indomethacin, G-TsF/TGF-beta-2 is the only suppressive factor released by the cultured glioblastoma cells, as indicated by neutralizing experiments with monospecific neutralizing antibodies (22, unpublished results).

THE GLIOBLASTOMA DERIVED T CELL SUPPRESSOR FACTOR, A MEMBER OF THE TRANSFORMING GROWTH FACTOR BETA GENE FAMILY

The peptide TGF-beta was originally characterized by its ability to phenotypically transform rat fibroblasts and is a homodimer with a molecular mass of 2.5×10^4 daltons. TGF-beta is a highly ubiquitous protein and has been purified from several normal tissues as well as from neoplastic sources. At least two forms of homodimeric TGF-beta have been described and have been named TGF-beta type 1 and type 2. The two peptides have apparently identical biological activities, although they differ in sequence and in the binding patterns to cellular receptors (for review see reference 24). Based on the results of the sequence determination (21), it is clear that G-TsF and TGF-beta-2 are synonymous terms for the same molecule. The functional data obtained with G-TsF/TGF-beta-2 may therefore also reflect the effects of TGF-beta-1.

In analogy to TGF-beta-1, G-TsF/TGF-beta-2 is translated as an inactive precursor protein from which the mature active peptide is released by proteolytic cleavage. The total length of 112 amino acids of the mature protein is the same as that of TGF-beta-1 and the two peptides share 71 % homology, while the first 302 amino acids of the predicted precursor molecule of G-TsF/TGF-beta-2 shows only partial homology (31 % identity) with the TGF-beta-1 precursor (21). All nine cysteins in the mature protein are conserved and biological activity depends on the presence of dimers of peptides.

IMMUNOLOGICAL ACTIVITIES OF G-TsF/TGF-BETA-2

Originally, the SN of glioblastoma cells was found to inhibit the ConA-induced proliferative response of thymocytes, the IL-2 or antigen induced growth of T cell lines, and the generation of cytotoxic T cells (15). These observations were confirmed when using purified G-TsF/TGF-beta-2 (23) or TGF-beta-1 (25). Interestingly, the inhibition of the thymocyte proliferation was completely neutralized by the addition of IL-2 and partially neutralized by TNFalpha. In the antigen dependent T cell assay using Ia-restricted ovalbumin-specific T cells (OVA-7), the factor was found not to interfere with recognitive interactions of antigen by the T cells nor with the initial triggering event following interaction of antigen with the antigen receptor. A direct cytotoxic effect of G-TsF/TGF-beta-2 on the OVA-7 T cells is unlikely as the initial release of IL-2 induced by antigen, or of IL-3 triggered by IL-2 was not impaired within the 24 hr period tested, and the viability of the T cells was not or only slightly affected by G-TsF/TGF-beta-2 (23). Moreover, the factor clearly inhibited also the growth-promoting effect of phorbol-ester (PMA) and calcium ionophore, which has been reported to substitute for antigen or lectin by stimulating both IL-2 receptor expression and secretion of IL-2 (24). The combined addition of two anti-IL-2 receptor antibodies (PC 61 and 7D4) completely blocked the proliferative response of the T cells to

IL-2 (25), however, only partially inhibited the PMA/ionophore effect on OVA-7 T cells. The PMA/ionophore-induced T cell activation pathway in the presence of anti-IL-2 receptor antibodies apparently is also inhibited by G-TsF/TGF-beta-2 (23). These data together suggest that rather than directly inhibiting IL-2-induced activation of IL-2 receptors, G-TsF/TGF-beta-2 may interfere with another T cell activation pathway common to IL-2 and PMA/ionophore stimulation. This pathway may also be influenced by macrophage-derived signals: TNFalpha had two opposite effects on OVA-7 T cells, a direct inhibition of the IL-2-induced proliferation, as well as a slight but significant reduction of the G-TsF/TGF-beta-2 suppression, which was not observed with IL-1beta, IL-3, IL-4, BSF-2/IL-6 or interferon-gamma (23).

Recently, attention has been focused on lymphokine activated killer (LAK) cells as a novel cell type found to exert tumor cell cytotoxicity (27). LAK cells are generated by culturing peripheral blood mononuclear cells with IL-2 for several days and lyse without MHC restriction both NK-resistant and NK-sensitive tumor cells in vitro. With regard to the potential role of LAK cells in the treatment of malignant tumors (28) including gliomas (29), the effect of TGF-beta on LAK cell generation and activity was studied by several investigators (30,32). Kuppner and coworkers found that the IL-2 induced generation of LAK cells from both normal blood donors as well as from glioblastoma patients are inhibited by G-TsF/TGF-beta-2 (30) and that the inhibition can be reduced by using higher concentrations of IL-2 (500 units/ml) during the induction phase. The suppressive effect of the factor is most significant during early stages of LAK cell generation and no inhibition is seen when G-TsF/TGF-beta-2 is added directly to the cytotoxicity assay (30). Comparable results were obtained in an independent investigation using human recombinant TGF-beta-1 and porcine TGF-beta-2 (32).

SUMMARY AND CONCLUSIONS

Impaired cell-mediated immunity has been well documented in patients with glioblastoma. It may well be that the suppressor activity detected in cyst fluid of glioblastoma as well as in the patient serum before but not after tumor removal is mediated, at least in part, by a soluble factor (G-TsF) recently identified and purified from culture medium of human glioblastoma cell lines, and for which cDNA clones were isolated. G-TsF is a member of the transforming growth factor beta (TGF-beta) gene family, structurally identical to TGF-beta-2. G-TsF/TGF-beta-2 inhibits mitogenic actions of IL-2 on T lymphocytes, the development of cytotoxic T cells (CTL) in mixed lymphocyte cultures, and the generation of lymphokine activated killer (LAK) cells, but not the cytotoxic effect of preformed CTL or LAK cells.

In malignant glioblastoma which express both MHC class I and class II molecules, macrophages/microglial cells and tumor infiltrating T cells have been identified. However, the T cells infiltrating the tumor tissue have a reduced proliferative response to mitogenic and allogeneic stimuli and exhibit low levels of cytolytic activity, compaired to autologous peripheral blood lymphocytes. This limited T cell activation may be due to secretory products of tumor cells such as G-TsF/TGF-beta-2.

ACKNOWLEDGEMENTS

This work has been supported by grants from the Swiss National Science Foundation (No. 3.930-0.87) and the Swiss Cancer League.

REFERENCES

1. Shirai Y (1921) Jpn Med World 1:14-21
2. Murphy JB, Sturm E (1923) J Exp Med 38:183-189
3. Medawar PB (1948) Br J Pathol 29:58-69
4. Weiner HL, Bhan AK, Burks J, Gilles F, Kerr C, Reinherz E, Hauser SL (1984) Neurobiology 34(Suppl):112-116
5. Ben-Nun A, Wekerle H, Cohen JR (1981) Eur J Immunol 11:195-199
6. Fontana A, Frei K, Bodmer S, Hofer E (1987) Immunological Reviews 100:185-201
7. Fontana A, Bodmer S, Frei K (1987) Lymphokines 14:91-121
8. Brooks WH, Netzley MC, Normansell DE, Horwitz DA (1972) J Exp Med 136:1631-1647
9. Kikuchi K, Neuwelt EA (1983) J Neurosurg 59:790-799
10. Brooks WH, Latta RB, Mahaley MS (1981) J Neurosurg 54:331-337
11. Coakham HB, Kornblith PL, Quinindley EY, Pollock LA, Wood WC, Hartnett LC (1980) JNCl 64:223-233
12. Dick SJ, Macchi B, Papazoglou S, Oldfield EH, Kornblith PL, Smith BH, Gately MV (1983) Science 220:739-742
13. Carrel S, de Tribolet N, Gross N (1982) Eur J Immunol 12:354-357
14. Fontana A, McAdam KPWJ, Kristensen F, Weber E (1983) Eur J Immunol 13:669-679
15. Fontana A, Hengartner H, de Tribolet N, Weber E (1984) J Immunol 132:1837-1844

16. Frei K, Bodmer S, Schwerdel C, Fontana A (1985) J Immunol 135: 4044-4047

17. Larsson I, Landstrom LE, Larner E, Lundgren E, Miorner H, Stannegard O (1978) Infection and Immunity 22:786-789

18. Lauro GM (1986) Acta Neuropathol 69:278-282

19. Yasukawa K, Hirano T, Watanabe Y, Muratani K, Matsuda T, Nakai S, Kishimoto T (1987) EMBO Journal 6:2939-2945

20. Wrann M, Bodmer S, de Martin R, Siepl C, Hofer-Warbinek R, Frei K, Hofer E, Fontana A (1987) EMBO Journal 6:1633-1636

21. de Martin R, Haendler B, Hofer-Warbinek R, Gaugitsch H, Wrann M, Schlüsener H, Seifert JM, Bodmer S, Fontana A, Hofer E (1987) EMBO Journal 6:3673-3677

22. Ikeda T, Lioubin MN, Marquardt H (1987) Biochemistry 26: 2406-2410

23. Siepl C, Bodmer S, Frei K, MacDonald HR, de Martin R, Hofer E, Fontana A (1988) Eur J Immunol 18:593-600

24. Sporn MB, Roberts AB, Wakefield LM, de Combrugghe B (1987) J Cell Biol 105:1039-1045

25. Lowenthal JW, Corthésy P, Tougne C, Lees RK, MacDonald HR, Nabholz M (1985) J Immunol 135:3988-3994

26. Ranges GE, Figari IS, Espevik T, Palladino MA (1987) J Exp Med 166:991-998

27. Grimm EA, Mazumber A, Zhang HZ, Rosenberg SA (1982) J Exp Med 155:1823-1841

28. Rosenberg SA, Lotze MT, Muul LM, Chang AE, Avis FP, Leitman S, Linehan WH, Robertson CN, Lee RE, Rubin JT, Seipp CA, Simpson CG, White DE (1987) New Engl J Med 316:889-897

29. Jacobs SK, Wilson DJ, Kornblith PL, Grimm EA (1986) Cancer Res 46:2101-2104

30. Kuppner MC, Hamou MF, Bodmer S, Fontana A, de Tribolet N (1988) Int J Cancer 42(4):1-6

31. Madisen L, Webb NR, Rose TM, Marquardt H, Ikeda T, Twardzik D, Seyedin S, Purchio AF (1988) DNA 7:1-8

32. Espevik T, Figari IS, Ranges GE, Palladino MA (1988) J Immunol 140: 2312-2316

Cerebral gliomas. G. Broggi and M.A. Gerosa, editors

TOXIN CONJUGATES FOR KILLING OF BRAIN TUMOR CELLS IN VITRO

COLOMBATTI M, DELL'ARCIPRETE L, BISCONTI M, STEVANONI G, *GEROSA MA and TRIDENTE G

Istituto di Scienze Immunologiche and *Dipartimento di Neurochirurgia,
University of Verona

INTRODUCTION

Gliomas are the most frequent primitive tumors of the human central nervous system. The prognosis associated with the most malignant form, glioblastoma multiforme, remains unfavourable in spite of surgery, radiotherapy and chemotherapy.

Histologically human glioblastomas are composed of heterogenous cell types. This heterogeneity, confirmed by phenotipic and genotypic studies of cultured glioma cell lines (1) appears to contribute to the variable response of this type of tumors to conventional chemotherapy or radiotherapy (2,3).

Monoclonal antibodies (MoAb) have opened new perspectives in cancer therapy because they specifically recognize tumor cells. Such MoAb can be used as carriers of potent toxins, thereby providing cell-specific cytotoxic agents (Immunotoxins, IT) that could selectively destroy neoplastic cells (for a review see ref. 4), permit manipulation of the immune system by selectively eliminating subsets of lymphocytes (5), or eradicate target cells from the bone marrow before allogeneic or autologous transplantation (6).

The toxin acts catalitically so it is much more potent than chemoterapeutic drugs that act stoichiometrically to block cellular functions. Toxins that are frequently used originate from plants (ricin, abrin) or bacteria (diphtheria toxin) and act by inhibiting protein synthesis (7). In this study we have used ricin purified from castor bean seeds of Ricinus communis. Ricin consists of two polypeptide chains with different functional activity: the A-chain is an enzyme that catalitically inactivates ribosomal protein synthesis whereas the B-chain binds to terminal residues of galactose present on cell surfaces and also facilitates A-chain penetration to the cytosol (7). One molecule of A-chain in the cytosol is sufficient to kill a cell (8).

MoAb can be chemically linked to intact toxin or to its A subunit (6). Comparison of ricin A-chain (RTA) conjugates with intact ricin conjugates has shown that intact ricin conjugates are more potent in killing target cells in vitro, but are less specific due to the non selective binding of the B-chain (6). In vitro, specificity of IT-ricin can be achieved by adding excess of lactose or galactose to the culture medium. Lactose competitively blocks the binding site of ricin B-chain (9). The in vitro activity of A-chain conjugates varies considerably depending on the cell surface antigen (Ag) to which the antibody is directed (10). Endocytosis and cytosol entry pathways may be involved in surface Ag mediated IT cytotoxicity and may explain the variable toxicity of RTA IT. Specific toxicity of RTA conjugates can be increased by the use of lysosomotropic substances (11). The mechanism of this potentiation is unknown. Amines and ionophores increase the pH in lysosomes and disrupt intracellular vesicle traffic (11), probably promoting A-chain translocation from endocytotic vesicles to the cytoplasm.

We have investigated the in vitro sensitivity of human glioma cells to conjugates synthesized using intact ricin or RTA and directed to different cell surface structures: a) an Ag expressed by human gliomas, GE2 (12) and b) the cell surface receptor for transferrin (13). We have used a MoAb anti GE2 and transferrin (Tfn) as toxin vehicles. Tfn receptors are expressed by growing tumor cells as well as by normal proliferating cells (14). Normal brain tissue expresses instead non significant levels of Tfn receptors as compared to neoplastic tissue (15). This raise the possibility of using Tfn receptor targetted conjugates for brain tumor therapy along with MoAb-toxin conjugates.

MATERIAL AND METHODS

Reagents

Ricin toxin was purified from castor beans as previously described (16). RTA was a kind gift of Dr. P. Casellas (Sanofi Recherche, Montpellier, France). Anti glioma MoAb GE2 was kindly supplied by Dr. S. Carrel (Ludwig Institute, Lausanne, Switzerland). Iron saturated Tfn was purchased from Miles. Ricin-MoAb conjugates were made with thioether linkage as described (17). RTA was linked to MoAb or to human Tfn by a disulphide bridge, as described previously (18). Toxin-MoAb and toxin-Tfn conjugates were purified by gel filtration on a TSK-3000 HPLC column.

Target cells

The Hu 126 cell line was established in our laboratory by primary explants of a tumor (glioblastoma multiforme) removed from a patient with an intracranial mass. Hu 126 cells were maintained in vitro by serial passages in RPMI 1640-10% FCS at 37°C in a humidified atmosphere of 5% CO_2, 95% air.

Protein synthesis assay

Protein synthesis was assayed by incubating $3x10^4$ cells in 90 µl leucine free RPMI-10% FCS in 96-well microtiter plates for 6 hr at 37°C. Protein synthesis was estimated by evaluating incorporation of ^{14}C-leucine in treated or control cells.

RESULTS

Cytotoxic effects of IT or Tfn-conjugates on human glioma cells

The cytotoxicity of anti GE2 and Tfn heteroconjugates was investigated in a protein synthesis inhibition assay. As illustrated in Tab. I, anti GE2-RTA in the absence or in the presence of 80 nM monensin killed 50% of the target cells at $1.5x10^{-8}$M and $8x10^{-9}$M respectively. Compared to the effect of RTA alone, the cytotoxicity of anti GE2-RTA was 13 times higher (Tab. I). GE2-ricin assayed in the presence of 100 mM lactose to reduce cell binding via ricin B-chain, inactivated 50% target cells at $2x10^{-10}$M, being about 1,000 times more cytotoxic than RTA. When compared to the cytotoxic effect of ricin in the presence of 100 mM lactose, however, its specific cytotoxicity was only about 15 fold higher (Tab. I). Binding of GE2-RTA and GE2-ricin to target cells was demonstrated by their higher toxicity as compared to ricin in the presence of lactose and also confirmed by cytofluorometry analysis. Consequently, the failure of anti GE2 IT to inactivate effectively target cells could not be attributed to cell surface antigen density. The ability of anti GE2 MoAb, used at saturating concentration, to induce modulation of the surface target Ag was then investigated. A variable decrease in the percentage of Ag^+ cells or in the cell surface Ag density is commonly observed when MoAb-induced modulation takes place (19). After 48 hr culture in the presence of an excess anti GE2 MoAb, the percentage of $GE2^+$ cells was 79.3%, a value comparable to that found in untreated control culture (82%). Thus, treatment of Hu 126 cells with anti GE2 MoAb did not result in GE2 Ag modulation.

The cytotoxic activity of RTA linked to human diferric Tfn was then investigated. Tfn-RTA showed undetectable toxicity up to 10^{-9}M in the absence of monensin, but was able to kill 50% target cells at $3x10^{-11}$M in the presence of 80 nM monensin, a 5,000 fold lower concentration than RTA in the presence of 80 nM monensin (Tab. I). Serum proteins, including free Tfn, are found within the cerebrospinal fluid (CSF), although in a lower concentration than in the serum (20). We therefore evaluated whether free Tfn present in the human CSF could reduce the effect of Tfn conjugates on human glioma cells. To this end, glioma target cells were treated with Tfn-RTA + monensin in the presence of 80% human pooled CSF. As shown in Tab. I, 80% CSF resulted in a 10 fold reduction of Tfn-RTA activity.

TABLE I

EFFECT OF HETEROCONJUGATE TREATMENT ON HU 126 HUMAN GLIOMA CELLS

Toxin or conjugate	IC50*	Specificity Factor**
RTA	2×10^{-7} M	1
RTA + monensin	1.5×10^{-7} M	1
Ricin	5×10^{-11} M	1
Ricin + lactose	3×10^{-9} M	1
Anti GE2-RTA	1.5×10^{-8} M	13
Anti GE2-RTA + monen.	8×10^{-9} M	18
Anti GE2-Ricin + lac.	2×10^{-10} M	15
Tfn-RTA	$>10^{-9}$ M	<200
Tfn-RTA + monen.	3×10^{-11} M	5,000
Tfn-RTA + monen. + CSF	3×10^{-10} M	500

* IC50 is the molar concentration of toxin or conjugate inhibiting 50% protein synthesis.

** The Specificity Factor was calculated according to the following formula: IC50 of toxin / IC50 of conjugate.

Kinetics of cellular protein synthesis inactivation by Tfn-RTA conjugates

The suitability of anti-tumor conjugates for in vivo application is determined also by their ability to kill high numbers of target cells within a short time. We have assayed the kinetics of Tfn-RTA at 10^{-9} M, a concentration well below the concentration at which RTA begins to be nonspecifically toxic for target cells. Tfn-RTA in the presence of monensin inactivates over 90% of Hu 126 cells within 5 hr.

To reproduce the conditions that may occur in an in vivo situation, 80% pooled CSF was included in the assay. Target cells were treated at a conjugate concentration (10^{-10}M), which is probably several fold lower than the concentration required to saturate the cell surface receptors for Tfn. As shown in fig. 1, Tfn-RTA + monensin killed about 80-90% target cells within 10-12 hr in the presence of 80% pooled CSF.

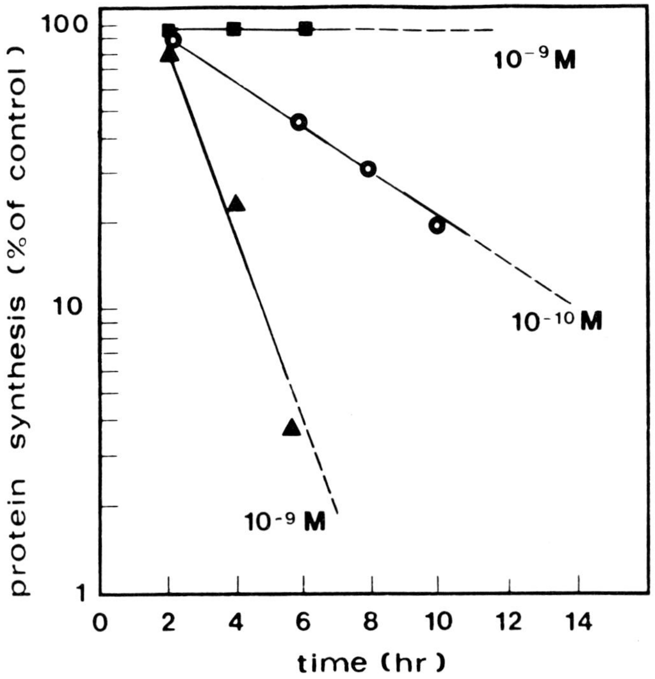

Fig. 1. Kinetics of protein synthesis inhibition by Tfn-RTA conjugate. Cells were treated with 10^{-9}M Tfn-RTA (■), 10^{-9}M Tfn-RTA + monensin 80 nM (▲) or 10^{-10}M Tfn-RTA + monensin 80 nM + 80% CSF (○) for the times shown in the figure. Results plotted as log protein synthesis versus time are expressed as percent control cultures.

DISCUSSION

Some properties of IT make them especially attractive as a complement to conventional therapy in the treatment of brain tumors: a) potent specific activity against target cells; b) cytotoxic activity not affected by oxygen tension, a factor often influencing the sensitivity of brain tumors to radio-chemiotherapy

(21, 22); c) cell killing independent from cellular cycle so that they are effective also against slow growing tumors.

We have evaluated the anti tumor potential of cytotoxic heteroconjugates directed to various cell surface structures expressed by human glioma cells.
IT to the GE2 Ag show low cytotoxic effects against target cells, in spite of the abundance of cell surface Ag available. High number of cell surface binding sites may not be sufficient to allow effective cell intoxication (23). Internalization and delivery of IT to an intracellular subsite that facilitates the passage of internalized toxin to the cytosol is required for IT high potency (6). GE2 Ag presumibly is inefficient to this respect. The Tfn receptor represents instead an optimal target for cytotoxic heteroconjugates. The Tfn receptor is a membrane spanning glycoprotein necessary for iron transport to the cytosol (13). Treatment of brain tumors with Tfn conjugates could be particularly advantageous, since iron is essential for tumor growth and it would be unlikely that target cells could escape Tfn conjugate mediated killing by modulation (24) or loss (25) of the Tfn receptor.

Use of lysosomotropic substances and ionophores greatly enhances the cytotoxic activity of RTA IT, possibly because of their effect on endosomal pH and on intracellular membrane organization. The cytotoxic activity of Tfn-RTA in the presence of the ionophore monensin was greatly increased in our experiments.
Kinetics of protein synthesis inhibition by toxins and IT is a function of receptors occupancy (9). We have observed fast cell inactivation at concentrations between 10^{-9}-10^{-10}M of Tfn-RTA. At the concentration of Tfn-RTA used in our study only a small fraction of the Tfn receptor are saturated. The kinetics of glioma cell inactivation by Tfn-RTA could be even faster than we reported. Techniques of conjugate delivery that allow persistence of high concentrations of Tfn-RTA in the site of the tumor may result in several logs of target cell killing in vivo. The presence of free Tfn competing with Tfn-RTA for binding at the cell surface did not reduce considerably the kinetics of cell killing. This could be ascribed to the rapid recycling (8-10 min.) of Tfn receptors following Tfn binding (26) and to the continuos availability of unoccupied receptors during treatment.

Some considerations should be made about the usefulness of our system: 1) isolated RTA is toxic for intact cells only at very high concentrations and several logs of specific killing could be obtained by linkage to Tfn; 2) iron loaded Tfn is the natural ligand of the Tfn receptor and its binding affinity is in the order of $10^8 M^{-1}$. Although the CNS is an "immunologically privileged" site (27) where immune reactions rarely take place under normal conditions, they might occur with a higher frequency during the course of a blood brain barrier breakdown, therefore use of human Tfn may limit the immune response to only one of the components of the conjugate.

Considering the properties of Tfn-RTA conjugates described in this article it could be predicted that in vivo application of this type of reagents could produce clinical results.

ACKNOWLEDGEMENTS

This work has been supported in part by PFO-CNR, Associazione Italiana per la Ricerca sul Cancro (A.I.R.C.), Regione Veneto and Ministero Pubblica Istruzione.

REFERENCES

1. Bigner DD, Bigner SH, Panten J, Wetermark B, Mahaley MS, Rueslathi E, Herschman H, Eug LF, Wikstand CJ (1981) J Neuropathol Exp Neurol 40:1983

2. Walker MD, Alexander E Jr, Hung WE, Mc Cathy CS, Mahaley MS, Mealey J, Nassel HA, Owens G, Ransohoff J, Wilson CB, Gehan EA, Strike TA (1978) J Neurosurg 49:333

3. Yung Wai-Kwan A, Shapiro WR (1982) Cancer Res 42:992

4. Vitetta ES, Fulton RJ, May RD, Till M, Uhr JW (1987) Science 238:1098

5. Vitetta ES, Krolick KA, Miyama-Inaba M, Cushley W, Uhr JW (1983) Science 219:644

6. Youle RJ, Colombatti M (1986) In: Roth JA (ed) Monoclonal Antibody in Cancer: Advances in Diagnosis and Treatment. Futura Publishing Company, New York, pp 173:213

7. Olsnes S, Pihl A (1982) In: Cohen P, van Heyningen S (eds) Molecular Action of Toxins and Viruses. Elsevier, Amsterdam, pp 51:105

8. Carrasco L, Fernandez-Puentes C, Vasquez D (1975) Eur J Biochel 54:499

9. Youle RJ, Neville DM Jr (1982) J Biol Chem 257:1598

10. Casellas P, Brown JP, Gros O, Gros P, Hellstrom I, Jansen FK, Poncelet P, Roncucci R, Vidal H, Hellstrom KE (1982) Int J Cancer 30:437

11. Casellas P, Bourrie BJ, Gros P, Jansen FK (1984) J Biol Chem 259:9359

12. Schnegg JF, Diserens AC, Carrel S, Accolla RS, De Tribolet N (1981) Cancer Res 41:1209

13. Hamilton TA, Wada HG, Sussman HH (1979) Proc Natl Acad Sci USA 76:6406

14. Gatter KC, Brown G, Trowbridge IS, Woolston RE, Mason DY (1983) J Clin Pathol 36:539

15. Zovickian J, Gray Johnson V, Youle RJ (1987) J Neurosurg 66:850

16. Colombatti M, Johnson VG, Skopicki HA, Fendley B, Lewis MS, Youle RJ (1987) J Immunol 138:3339

17. Youle RJ, Neville DM Jr (1980) Proc Natl Acad Sci USA 77:5483

18. Colombatti M, Nabholz M, Gros O, Bron C (1983) J Immunol 131:3091

19. Chatenoud L, Bach JF (1984) Immunol Today 5:20

20. Walsh MJ, Tourtellotte WW (1984) J Neurochem 43:1278

21. Hall EJ (1982) Int J Radiat Oncol Biol Phys 8:373

22. McNally NJ (1982) Int J Radiat Oncol Biol Phys 8:593

23. Press OW, Vitetta ES, Farr AG, Hansen JA, Martin PJ (1986) Immunol 102:10

24. Old LJ, Stockert E, Boyse EA, Kim JH (1968) J Exp Med 127:523

25. Hyman R, Cunningham K, Stallings V (1980) Immunogenetics 10:261

26. Dautry-Varsat A, Ciechanover A, Lodish H (1983) Proc Natl Acad Sci USA 80:2258

27. Scheinberg LC, Ortmeyer AE, Sussman HH (1968) Progr Neurol Surg 2:267

EPIDEMIOLOGY

Cerebral gliomas. G. Broggi and M.A. Gerosa, editors

SURVEY ON RISK FACTORS OF CEREBRAL GLIOMA: A CASE-CONTROL STUDY

F. MENEGHINI**, S. MINGRINO, P. ZAMPIERI, G. SOATTIN, P. LONGATTI, L. CASENTINI, M. GEROSA, C. LICATA, M.C. ZOPPETTI, F. GRIGOLETTO*

Neurosurgical Divisions of Padova, Treviso, Verona, Vicenza (Italy)
*Department of Statistical Sciences University of Padova (Italy)
** Department of Biostatistics and Epidemiology, Fidia Research Laboratories, Abano Terme - Padova (Italy)

Case-control studies, also commonly called retrospective studies, belong to the analytical epidemiology, a branch of epidemiology which deals with the possible association between the onset of a pathology and the occurence of particular conditions. The main purpose of these studies is to test and develope etiologic hypotheses by means of the delucidation of cause-and-effect relationships (1).

In the case-control design, subjects with a particular disease, called cases, are selected for a comparison with a group of individuals in whom the condition of disease is absent (the controls).

Cases and controls are compared with respect to the presence of attributes considered to be relevant to the development of the disease. The measure of association between factors and disease is defined as Odds Ratio, the ratio of the proportion exposed-unexposed (Odds of exposure), in the case group, to the same proportion in the control group. This rate represents how many times more (or less), likely a disease occurs in the exposed as compared with the unexposed (2). For the index greater than one, a "positive" association is said to exist, for values less than one the association is "negative". Of course, to confirm or refute a result, that is a relationship of cause and effect, collateral evidence and biological plausibility must be provided.

In planning and conducting case-control studies, many sources of bias can intervene suggesting misleading associations. Biases can occur in defining cases (improper diagnosis) or controls (biased selection of study subjects) or in determining the past events. Moreover misleading conclusion can be due to the presence of confounding variables, that are extraneous parameters associated with both the disease and the study factor, without being consequence of exposure (3).

The case-control method of investigation is based on information of difficult finding and complex validation. Moreover this methodology is relatively unfamiliar to medical community and difficult to explain. Although these disadvantages, such a technique is the most appropriate for rare or long latency diseases, relatively inexpensive and quick to conduct. Finally this research strategy is used when initiating an explanatory study of etiology, and multiple hypotheses are selected for investigation.

The SMEG Group (Multicentric Study of Epidemiology of Glioma) is conducting two surveys on cerebral glioma: a longitudinal and a case-control study. The study group is formed by the Neurosurgical Divisions of Padova, Verona, Treviso, Vicenza Hospitals and the Department of Biostatistics and Epidemiology of Fidia Research Laboratories, Abano Terme, with grants of Veneto Region. The design and some preliminary results of the case-control study are here reported.

The numerous studies on this pathology do not provide consolidated findings relatively to its etiology: therefore the aim of the study is essentially exploratory. A previous analysis (4) has been conducted to point out the surveyed risk factors, they are: familial aggregation, blood group, head trauma, history of cigarette and alcohol consumption, occupation, exposure to radiations, neurological and infective pathologies in the subject or in his relatives, drugs.

Cases are patients aged between 18 and 74 years, admitted to the Neurosurgical Divisions participating in the study, with diagnosis of glioma (oligodendroglioma, astrocytoma, glioblastoma, ependymoma, medulloblastoma) histologically proved at the time of admission. In order to avoid mistakes in defining cases all the diagnoses were reviewed by a senior neuropathologist.

Control subjects are patients admitted to the same hospitals of the cases with any condition other than glioma and without family tie with cases. They are singly matched to a case according to the following variables: sex, age (± 5 years), area of residence and time of hospitalization.

All the information is collected by a structured questionnaire designed in order to minimize reporting and coding errors.

An other survey conducted on patients of the participating Neurosurgical divisions revealed that about 70 percent of cases suffered of mental disorders. For this reason the questionnaire is administered to a relative of the case (next of kin) and, for sake of homogeneity, of the control. The interviewers are carefully trained personnel chosen among nurses

Up to now 171 case-control pairs out of the planned 270 were recruited. Table I reports the distribution by sex and age of the cases. It shows a higher frequency of males aged between 51 and 60, according to the findings reported by literature.

TABLE I

SEX AND AGE DISTRIBUTION OF 171 RECRUITED CASES

Age	Males No.	Males %	Females No.	Females %
< 30	13	11.9	9	14.5
31 - 40	11	10.1	10	16.1
41 - 50	21	19.3	8	12.9
51 - 60	34	31.2	18	29.1
61 - 70	23	21.1	14	22.6
> 70	7	6.4	3	4.8
All	109	100.0	62	100.0

To verify the representativeness of cases and test the possible presence of selection bias, we compared the characteristics of cases with those of the wider set of patients with glioma, recruited for the longitudinal study. In that study the recruitment procedure can be considered as exhaustive of the case population referring to the hospital. Tables II and III show the similarity of the two distribution by age and type of diagnosis.

TABLE II

PERCENTAGE DISTRIBUTION BY AGE OF THE CASES RECRUITED IN THE CASE-CONTROL AND LONGITUDINAL STUDIES

Age	Study: Case-control	Study: Longitudinal
< 30	12.9	13.6
31 - 40	12.3	10.7
41 - 50	17.0	15.7
51 - 60	30.4	33.8
61 - 70	21.6	21.6
> 70	5.8	4.6
All	100.0	100.0

TABLE III

PERCENTAGE DISTRIBUTION BY DIAGNOSIS OF THE CASES RECRUITED IN THE CASE-CONTROL AND LONGITUDINAL STUDIES

Diagnosis	Case-control	Longitudinal
Oligod.	5.9	8.0
Astroc. 1 2	21.8	20.6
Astroc. 3 4	67.6	65.8
Ependim.	3.5	4.1
Medullobl.	1.2	1.5
All	100.0	100.0

A further control of data quality was worked out in order to measure the level of bias induced by the next of kin. For about 10% of the control group a direct interview was carried out for the control. The minimal percentages of agreement (minimal with respect to all questions) between the two sources of information, control and his next of kin, was calculated. A percentage of agreement equal to 93.7 was observed as regards the presence of the factor (e.g. "Have you ever smoked?") and equal to 71.8 for the level of exposure (e.g. "How many cigarettes a day?"). For both the kinds of questions, the level of agreement was judged as satisfactory.

Finally Table IV reports some Odds Ratios and their 95% confidence intervals, to qualify the estimates.

TABLE IV

PRELIMINARY DATA: ODDS RATIO FOR SOME RISK FACTORS (171 PAIRS)

Factor	Odds Ratio	95% CI*
Head trauma	1.08	0.60 - 1.94
Blood group 0	1.60	0.99 - 2.61
Relatives with nervous disease	1.17	0.58 - 2.39
Relatives with tumor	1.18	0.75 - 1.85

* 95% confidence interval for Odds Ratio.

Nevertheless, because of the limited sample size, these values are to be considered as very early and possible to be modified by new data: therefore they are to be read as merely suggestive.

REFERENCES

1. Lilienfeld AM, Lilienfled DE (1980) Foundations of epidemiology. Oxford University Press, New York
2. Fleiss J.L. (1973) Statistical methods for rates and proportions. John Wiley and Sons, New York
3. Schlesselman JJ (1982) Case-control studies. Oxford University Press, New York
4. Mingrino S, Zampieri P, Gallato R, Di Stefano E, Gerosa M, Nicolato A, Casentini L (1985) Review of Epidemiological Studies on Cerebral Glioma and Presentation of a New Cooperative Study in the Veneto Region. Advances in the Biosciences 58: 215-221.

Cerebral gliomas. G. Broggi and M.A. Gerosa, editors

EPIDEMIOLOGY OF CEREBRAL GLIOMA: A MULTI-CENTRE STUDY IN THE VENETO REGION OF ITALY

S. MINGRINO, P. ZAMPIERI, G. SOATTIN, A. PADOAN, M. GEROSA, C. LICATA, M.C. ZOPPETTI, L. CASENTINI, U. FORNEZZA, P. LONGATTI, F. MENEGHINI*

Neurosurgical Divisions of Padova, Treviso, Verona, Vicenza (Italy)
* Department of Biostatistics and Epidemiology, Fidia Research Laboratories, Abano Terme - Padova (Italy)

INTRODUCTION

The tendency of modern clinical studies in oncology is to overcome the personal physician/patient relationship and extend the limited experience of the single centre. It is increasingly important for the physician to obtain precise data as to the frequency, distribution and behaviour of tumoral disease in large areas and this has given rise to the creation of cancer registries (1).

A valid cancer registry should contain the elements of the population-based registry, to express descriptive epidemiology parameters (incidence, prevalence, sex and age distribution, etc.) and those of the hospital-based registry (clinical findings, diagnostic procedures, therapeutic choices and outcome of treatment). Such a registry is highly complex to organize and can only stem from a definite project launched by the political authorities on a regional or even a national level. No such registry exists in Italy. The available epidemiological data, published by the Central Institute of Statistics, regard mortality and are classified on the basis of very general criteria (e.g. benign vs malignant tumors) of marginal clinical interest. There are some population-based registries regarding limited areas. The world directory of cancer registries, published by the IARC of Lyon (2), mentions the two cancer registries of Lombardy and Tuscany and a population-based study on tumors of the CNS was conducted in the province of Trento a few years ago. In our own region (the Veneto) a proposal for the creation a Regional Tumor Registry on the lines of the Swedish experience (4) is currently under discussion. Meanwhile a Veneto regional study on brain tumors has been planned with a view to collecting elements of both descriptive epidemiology and clinical management.

DESCRIPTION OF THE VENETO STUDY

Our research on the epidemiology of glioma includes a longitudinal study as well as a case-control study on risk factors (discussed separately) with all four regional Neurosurgical Centres (Padova, Treviso, Verona and Vicenza) taking part. Thus the majority of cases of brain tumor occurring in the Region have been considered, as these patients are sent to the Neurosurgeon for consultation or surgery. Clearly, the study has not included any cases hospitalized outside the Region or in private institutes and any cases never seen by a Neurosurgeon: considering the kind of service available and the sanitary habits of the population concerned, such cases must be a very small minority.

The study began in June 1985 and cases were collected for two years. The study is still underway with regard to data quality checks and the completion of patients' follow-up data.

The group of neurosurgeons involved in the research at each Centre has remained substantially unchanged. Data collection was based on a pre-set protocol but each Centre made independent diagnostic and therapeutic decisions.

All histological findings were re-examined by the same neuropathologist - Prof. F. Gullotta of Munster (FRG) - to guarantee uniform classification of the histological type of tumor.

The study collected 571 cases of primary brain tumor, i.e. WHO-classified neuro-epithelial tumors (astrocytomas, oligodendrogliomas, ependymomas, choroid plexus tumors, pineal cell tumors, neuronal tumors such as ganglio-cytomas and poorly-differentiated tumors such as glioblastomas and medulloblastomas).

Of the 571 patients included in the study, 346 (60.6%) were from the Veneto and 225 (39.4%) from other regions. This is a useful indicator for those responsible for planning health services and it also enables a rough estimate of the incidence of primary brain tumors in our Region. The incidence of cerebral gliomas in various other studies is illustrated below.

Incidence of cerebral glioma

Barker, 1976 (5)	3.9/100,000/yr
Walker, 1985 (6)	3.9/100,000/yr
Ferrari, 1986 (3)	4.2/100,000/yr

This means that the annual incidence is of nearly 4 new cases per 100,000 people. The incidence in the Veneto region calculated from our data is 3.74/100,000/yr (4.38 for males and 3.13 for females). This result cannot be considered as final because some patients probably escaped our attention, but its similarity to the results of the population-based registries is indicative of the extensive recruitment of cases and it may be taken as a reliable minimum value.

As regards sex, the 571 cases considered comprised 331 males (57.9%) and 240 females (42.1%).

Age distribution was as follows:

- 41 cases aged < 10 yrs (7.2%)
- 43 cases aged 10-20 yrs (7.5%)
- 55 cases aged 21-30 yrs (9.6%)
- 62 cases aged 31-40 yrs (10.9%)
- 76 cases aged 41-50 yrs (13.3%)
- 151 cases aged 51-60 yrs (26.5%)
- 109 cases aged 61-70 yrs (19.1%)
- 34 cases aged > 70 yrs (5.9%)

The general data regarding therapy are summarized below

Type of treatment

	Yes	No	Unknown
Surgery	84.2%	15.8%	-
Radiotherapy	49.5%	49.1%	3.4%
Chemotherapy	5.0%	92.2%	2.8%

This shows a clear tendency in favour of surgery, radiotherapy is recommended fairly frequently and chemotherapy is seldom used. As regards the type of surgery, macroscopic total removal was performed in 173 cases (35.9%), subtotal extirpation was done in 148 cases (30.7%), partial removal in 59 (12.2%) and biopsy in 101 (20.9%). Operative mortality (from day of operation and up to 30 days) proved to be 5.6% (27 cases out of 481 operated). These 481 histologically-verified cases comprised: 111 low-grade astrocytomas, 143 anaplastic astrocytomas, 119 glioblastomas, 36 oligodendrogliomas, 19 medulloblastomas, 20 ependymomas, 33 others. In synthesis about 50% were cases of malignant astrocytoma, 25% of low-grade astrocytoma and 25% of the other less frequent tumors.

As previously mentioned, the study is still underway with regard to collecting patient's follow-up information so the final results are not yet available, though some interesting preliminary data are emerging. For example, the survival rate for our cases of malignant supra-tentorial astrocytoma is 42% at 12 months and 7% at 24 months. These values are very similar to those reported by Salcman (7) on 1561 cases of glioblastoma treated by means of surgery, radiotherapy and/or conventional chemotherapy collected from the literature.

REFERENCES

1. Muir CS, Demaret E, Boyle P (1985) In: Parkin DM, Wagner G, Muir CS (eds) The role of the registry in cancer control. IARC, Lyon, pp 13-26

2. Menck HR, Parkin DM (eds) (1986) Directory of computer systems used in cancer registries. IARC, Lyon

3. Lovaste MG, Ferrari G, Rossi G (1986) Epidemiology of primary intracranial neoplasms. Experiment in the Province of Trento (Italy), 1977-1984. Neuroepidemiology 5: 220-232

4. Moller TR (1985) In: Parkin DM, Wagner G, Muir CS (eds) The role of the registry in cancer control. IARC, Lyon, pp 109-119

5. Barker DJP, Weller RO, Garfield JS (1976) Epidemiology of primary tumors of the brain and spinal cord: a regional survey in southern England.

Journal of Neurology, Neurosurgery and Psychiatry 39: 290-296

6. Walker AE, Robins M, Weinfeld FD (1985) Epidemiology of brain tumors: the national survey of intracranial neoplasms. Neurology 35: 219-226

7. Salcman M (1985) The morbidity and mortality of Brain Tumors. A perspective on recent advances in therapy. Neurologic Clinics 2: 229-257

DIAGNOSIS

© 1989 Elsevier Science Publishers B.V. (Biomedical Division)
Cerebral gliomas. G. Broggi and M.A. Gerosa, editors

POTENTIAL LIMITS OF CT SCAN AND NMR

Passerini A., Strada L., Sberna M., Grisoli M.

Istituto Neurologico "C. Besta" - Milano

Supported by CNR N° 87.01567.44

Magnetic Resonance Imaging (MRI) has had a fulmineous temporal progression from the experimental prototype to the present sophisticated instrumentation. Very few years have passed from the introduction of MR and the excellent diagnostic capability recently demonstrated by this procedure. The quick progression of MRI can be sharply contrasted to the more protracted time that characterized the development of X-ray Computed Tomography (CT). Now, the current generation of CT scanners is the results of more than a decade of technologic improvement . In comparison with that , the evolution of MRI as been much more rapid . Presently, as MR is proved to be an efficacious tool for the diagnosis of neurologic diseases, increasing attention is being focused on the limitation of this procedure. In fact, MR image quality depend from several factors: signal-to-noise ratio (S/N R), contrast resolution, spatial resolution and artifact production. In early papers (1), MR was favourably welcome for the elimination of the artifacts - mainly bone artifacts - well known in CT imaging . Afterwards, many Authors pointed out that MRI can be disturbed by other types of artifacts (2) (fig. 1): a MRI artifact can be defined as any signal intensity, or void, wich does not have an anatomic basis (3). Motion-related signal alterations, comprehending not only gross motion but also intravascular flow effects and cerebrospinal-fluid (CSF) motion effects, chemical shift artifacts and ferromagnetic artifacts can significantly modify the MR image. Some artifacts, such as motion and the so called aliasing artifact can be easily manipulated. Other, such as RF pulse inhomogeneity, are more sporadic and difficult to diagnose, and correcting them will require field service engineers (3).

Both CT and MR produce cross-sectional, digital images of the body. Digital images are made of numbers arranged in a matrix: the density of the matrix must be enough to compose an image, when shades of gray are assigned to the numbers. The numberical value are an expression of the physical properties of the tissue; hence the final image gives information about these properties. CT imaging provides information about only one physical property, i.e. the attenuation of the X-ray beam due to the electron density of the tissue. MR imaging encodes different physical parameters: because of the complessity of its input, it is superior to CT in the depiction of the tissue. To expose these multifactorial characteristics of MR imaging, we will discuss some basic aspects.

MR imaging arises from the effect of strong magnetic field on the hydrogen nuclei (single protons) and the controlled perturbations of these nuclei by radiofrequency (RF) waves. The protons excited by RF waves emit in return a radiowave signal, the intensity of which is proportional to the number of protons. But the hydrogen density is not the only physical property of the tissue elucidated with MR imaging: MR also gives information about the protons environment and the state of their motion. T1 relaxation reflects the interaction of the hydrogen nuclei with molecular environment. T2 relaxation is related to the magnetic homogeneity of their environnement. Other physical properties of the tissue elucidated by MR are the rate of motion of the protons (flow, diffusion) and the type of molecule which surrounds the nuclei (chemical shift, magnetic susceptibility). Therefore, MR information are much more complex and multifactorial than CT one: as a consequence of this phenomenon, a lot of parameters are operator-dependent in each set of image acquisitions. One of the most important parameters which has to be selected during the examination is the type of RF pulse sequence, which is a time-dependent factor (5). The most commonly used pulse sequence is the Spin-Echo sequence (SE) (fig. 2), in which a 90° pulse rotates the magnetization vector 90° from the magnetic field; after that, a 180° pulse restores coherence. Another pulse sequence used in MR imaging is Inversion Recovery (IR), in which a 180° pulse followed by a 90° pulse is applied at regular intervals. Saturation Recovery (SR) is another pulse sequence initially introduced but now unfrequently employed in clinical practice.

In practical work there always is a conflict between image quality and scan time, and this is a very critical point in Paediatric Neuroradiology. On the purpose to cut examination time the operator can manipulate different parameters, such as the numbers of averages and the matrix. For the same reasons, in order to arrange the most appropriate pulse sequence for the particular problem the operator have to face, different techniques have been settled after invertion recovery, saturation recovery and spin-echo. Some of these techniques have as the main target the shortening of the examination time.

In 1983, Crooks et al. (4) proposed a new technique, the so called interleaved multislice technique, which allows a significant reduction of the examination time (about 10 times shorter when long TR is employed) (see fig. 1). Because the time in which a sequence of irradiation and data acquisition is performed is considerably shorter than TR, it is possible to repeat the imaging sequence on another section. In this manner, while waiting for the TR interval to elapse, we can obtain a number of images in the time just one image would be acquired with the single slice technique. Therefore, this technique allows a simultaneous acquisition on the whole region the operator has to study, i.e. the whole brain. Furthermore, this multisection procedure speeds up the examination time without damaging the image quality, i.e. tissue contrast, S/N R and spatial resolution.

Another technical device the operator can choose to perform the most appropriate examination for a particular clinical problem is the multiple (more than two) spin-echo (MSE) technique (fig. 3). In this case, the imaging gives information about a single section, but with a set of images at different echo-times. This is a time-consuming technique which is performed for a more complete characterization of the T2 structure of a lesion. Usually, in our experience, TR is the same as for the routine SE sequence (about 2000 msec); TRs are 30, 60, 90, 120 msec in the majority of cases. When the lesion we have to study is characterized by high T2 values, MSE technique shows hyperintensity at the last echos and enhances the contrast between the lesion and the surrounding structures: in this manner, a better detection of the lesion is allowed.

Another technical procedure which provides thin contiguous sections, sharp anatomic details and excellent contrast resolution is the three dimentional volume study (3D) (fig. 4). The difference between this technique and the other over mentioned is that not only a section but an entire volume is excited. Not differently than MSE, 3D is a time consuming procedure; however, it significantly improves the image quality. For this reason, it is expecially employed to study small anatomic regions, i.e. pituitary region, optic nerves and optic chiasm. Recently, fast imaging techniques with partial flip angles have been settled (6).

The first of them was the so called FLASH (fast low angle shot) sequence.Afterwards, a number of other were described: FFE (fast field echos), GRASS (gradient recalled acquisition in the steady state), FAST (fourier acquired steady state technique), FISP (fast imaging with steady procession) (fig. 5). All these techniques use a partial flip angle of the magnetization vector, short TRs and do not use 180° RF pulses in spin-echo refocusing. Instead, a gradient reversal technique is used to restore coherence. This partial flip MR imaging is a time-saving technique able to show additional information in short time. These techniques do not replace conventional spin-echo sequences, which presently are the basic of MR imaging. Otherwise, they are interesting for specific application: in the brain, the most important role of partial flip MR imaging is the detection of intracranial hemorrage; it can also be used in studying cerebrovascular flow phenomena. In the spine, the main application of these techniques is the rapid production of MR myelograms.

Till now, different devices to manipulate in a noninvasive manner MR image contrast have been briefly summarized. The paramagnetic contrast media such as Gadolinium-DTPA (Gd-DTPA) are a different approach to the same problem: to enhance a lesion (fig. 6). When an active blood-brain-barrier lesion is present, Gd-DTPA enhanced T1 weighted MR images show the lesion. So the contrast agent improves the ease of identification of a lesion and speeds up examination time. The primary application of the paramagnetic agents is not in Pediatric Neuroradiology because of their toxicity; at any rate they may

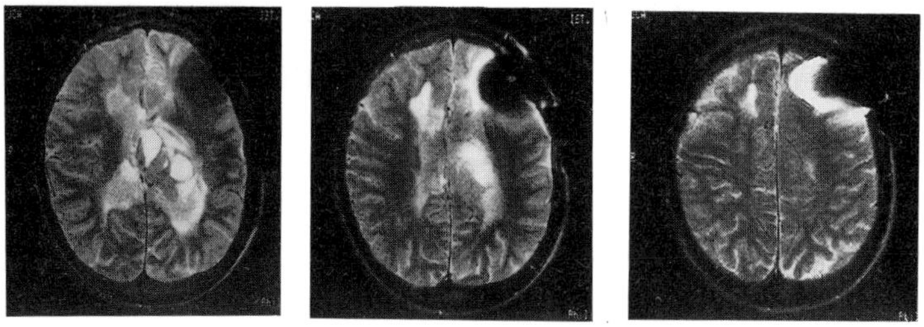

Fig. 1
Hydrocephalus due to a multicystic tumor in the pineal region. Shunt device in left frontal region. SE 1933/100 multislice technique. See the artifacts due to shunt catheter and metal objects.

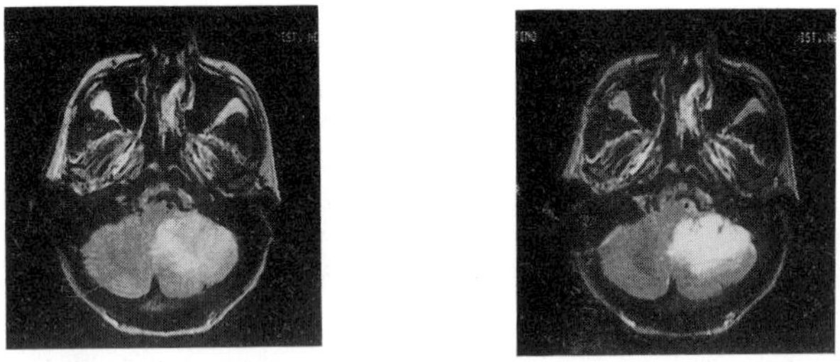

(a) (b)

Fig. 2
Posterior fossa metastatic lesion on SE double echo sequence. Images obtained with 8 mm section at 0.5 T. (a) SE 1933/50 (TR msec/TE msec).
(b) SE 1933/100 (TR msec/TE msec).

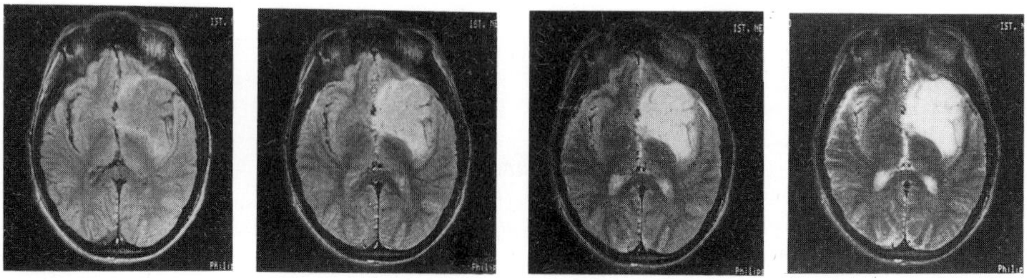

Fig. 3
Left fronto-temporal astrocytoma. Multiple spin-echo technique. SE 1800/30, 60, 90, 120 (TR msec/TE msec). Images obtained with 8 mm section at 0.5 T. In the last echos the hyperintensity of the lesion enhances the contrast between the lesion and the surrounding structures.

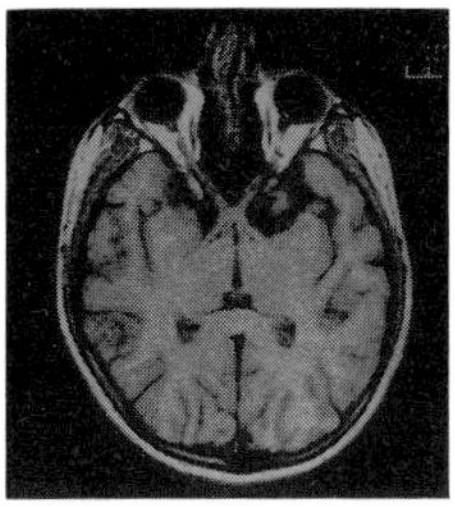

Fig. 4
Normal anatomic study. Three dimensional volume study (3D) of optic nerves and optic chiasm. Images obtained at 0.5 T., SE 300/30 (TR msec/TE msec), with a slice thickness of 3 mm: see spatial resolution and anatomic details.

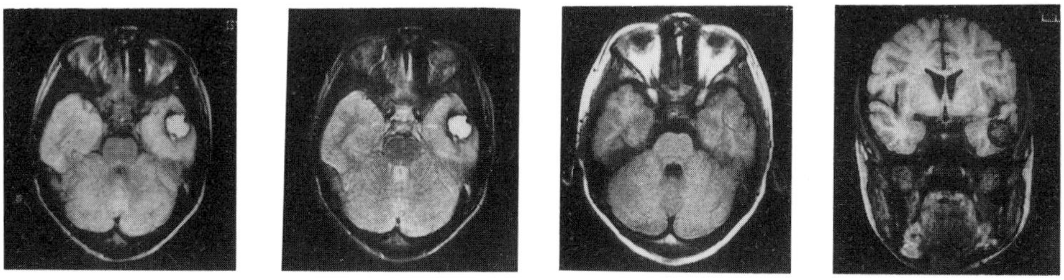

Fig. 5
Left temporal cavernous hemangioma. (a, b) SE 1933/50, 100 (TR msec/TE msec). Slice thickness 8 mm at 0.5 T. (c) SE 450/30 (TR msec/TE msec) with 8 mm section at 0.5 T. (d) Partial flip angle technique (FFE) obtained at 0.5 T., 410/20 (TR msec/TE msec). The partial flip procedure is superior to conventional T1 weighted SE image in demonstrating the hemorrhagic aspects of the lesion.

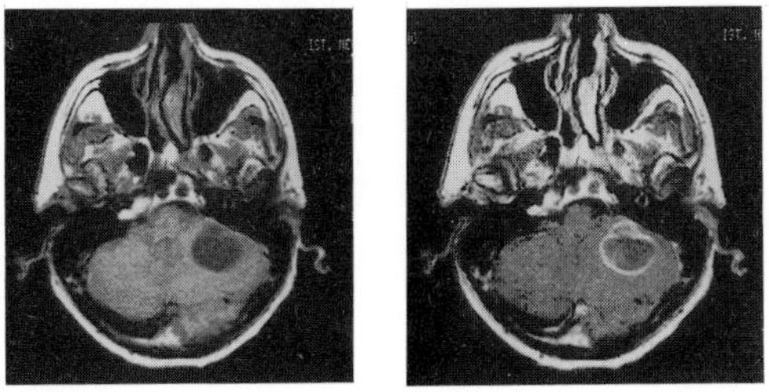

Fig. 6
The same case as in Fig. 1. (a) SE 500:30 (TR msec/TE msec). (b) SE 450/30 (TR msec/TE msec) 30' after Gd-DTPA injection. The comparison definitely shows the difference between basic T1 weighted MR images and Gd-DTPA enhanced T1 weighted images.

be considered as a supplemental tool for the diagnosis (7).

As a conclusion of this brief summary of the MR capabilities, it may be interesting to draw a comparison between CT and MR procedures. At the beginning, MR was a complemental examination and CT was a noninvasive alternative, wich had already reached chinical maturity. At present, MRI is sufficient for the diagnosis in several cases and even the examination time can be comparable to that or X-ray CT.

REFERENCES.

1. Brant-Zawadzki M, Davis PL, Crooks DE et al., NMR demonstration ofcerebral abnormaties: comparison with CT. AJNR 1983, 4: 117-124.
2. Kelly WM, Image artifacts and technical limitation. In: Magnetic Resonance Imaging of the Central Nervous System, Brant-Zawadzki M., Norman D. (eds) Raven Press, New York, 1987, pp. 43-82.
3. Harris R, Wesbey G, Artifacts in Magnetic Resonance Imagine, In Magnetic Resonance Annual 1988, Kressel HJ (eds) Raven Press, New York, 1988.
4. Crooks LE, Ortendahl DA, Kaufman L et al., Clinical efficiency of Nuclear Magnetic Resonance Imaging, Radiology 1983, 146: 123-128.
5. Brant-Zawadzki M, Norman D, Newton T et al., Magnete Resonance of the Brain: the optimal screening technique, Radiology 1984, 152: 71-77.
6. Winkler ML, Ortendahl DA, Mills TC et al., Characteristics of Partial Flip Angle and Gradient Reversal MR Imaging, Radiology 1988, 166: 17-26.
7. Brant-Zawadzki M, Berry I, Osaki L et al., Gd-DTPA in Clinical MR of the Brain: 1. Intraaxial Lesions, AJNR, 1986, 7: 781-788.

DIAGNOSTIC VALUE OF MAGNETIC RESONANCE IMAGING AND SPECTROSCOPY IN BRAIN TUMORS

William FEINDEL, Yvon ROBITAILLE, Douglas ARNOLD, Eric SHOUBRIDGE, Joseph EMRICH, and Jean-Guy VILLEMURE
McConnell Brain Imaging Center, Montreal Neurological Institute and Hospital, Montreal, Canada.

INTRODUCTION

Malignant cerebral gliomas have a uniformly poor prognosis. Recent advances in our understanding of their biology, though significant, have not yet been translated into cures. Much needs to be done to attain more than the palliative treatment now available. The hourly striking of the church clock of San Giacomo, next to the Villa Durazzo where we meet, serves to remind us, "For whom does the bell toll?". Certainly it tolls for our patient with malignant gliomas and, in a real sense, for all of us searching to develop more effective treatment for these relentless brain lesions.

This report highlights two new uses of nuclear magnetic resonance in the clinical management of brain tumors. First, it points out the special diagnostic advantages of magnetic resonance imaging (MRI) for early and exact detection of small structural lesions, exemplified here by gliomas and angiomas related to temporal lobe epilepsy. Secondly, it summarizes preliminary findings from the application of phosphorus-31 magnetic resonance spectroscopy (PMRS) for metabolic analysis applied to differential diagnosis of intracranial tumors.

About five hundred years ago in Milano, a historical contribution to imaging of the brain's anatomy was made by Leonardo da Vinci. He depicted for the first time the shape of the cerebral ventricles displayed by injecting melted wax into the cavities of the ox brain (Fig.1). MRI of an ox head shows the lateral and third ventricles on coronal views to update Leonardo's work (Fig.2).

The basic scheme of the three main methods of brain imaging in current use, include a data acquisition system, and a data transfer system that carries this information into the computer programmed to reconstruct images displayed on a video screen. In computerized axial tomography (CAT) the acquisition of energy data is made by an x-ray source activating a crystal gamma-ray detector. In positron emission tomography (PET) the positron emitting source of tracer in the brain replaces the external x-ray source and the number of gamma detectors is greatly multiplied to provide reconstruction of a chemical image of the brain. In MRI, the body is enveloped by the powerful magnetic field of the instrument and radio pulses perturb the chemical nuclei to produce a series of signals that can be received and measured externally. These signals are then processed by the computer system to produce a detailed anatomical image (1).

MRI for detecting tumors in temporal lobe epilepsy

Of some 100 operations each year for patients with focal epilepsy at the Montreal Neurological Institute, about 75 are for seizures involving the temporal lobe. In these, MRI identifies small structural lesions about 30% more effectively than does computerized axial tomography (CT). In reviewing 48 patients treated surgically for focal temporal epilepsy, MRI demonstrated twelve structural lesions confirmed by surgical pathology (2). Most of these were located in the mesial and anterior part of the temporal lobe. One third of these lesions were missed on CT, because the temporal lobe is inadequately visualized, due partly to the high electronic density of the bone of the skull base as compared to brain.

Before the advent of brain imaging, Mathieson (3) reviewed over 800 patients who had undergone surgical treatment for temporal lobe epilepsy; he found that about 1 in 5 harbored a structural lesion, such as a small tumor, angioma or hamartoma. In the past three years, MRI has increased the detection of such lesions to about 1 in 4 in our surgical series. In 70% of the remaining patients of this recent series, a definite increase of signal intensity appeared in the mesial temporal region, in many cases associated with astrocytic gliosis in the amygdala (4). Several case reports will illustrate these points.

1. Case F A man 36 years of age with seizures for over 30 years involving automatism, showed on MRI a discrete lesion with a high signal intensity in the amygdala and anterior hippocampus. Multi-planar views added details of the position and size of the lesion next to the temporal horn and behind the middle cerebral arterial stem (Fig.3). Pathological diagnosis was subependymoma.

2. Case M This man of 44 with seizures for 10 years had a ganglioglioma that gave on MRI signal intensity changes involving the hippocampus and amygdala (Fig.4).

3. Case Q A woman 29 years of age had olfactory seizures for four years that were related to a small angioma in the amygdala (Fig.5). The image was characteristic, with a "target" appearance having a center of increased signal intensity surrounded by a hypodense ring.

These examples of small lesions readily identified by MRI epitomize the correlation of clinical seizure pattern, focal EEG localization, operative findings and microscopic pathology in the

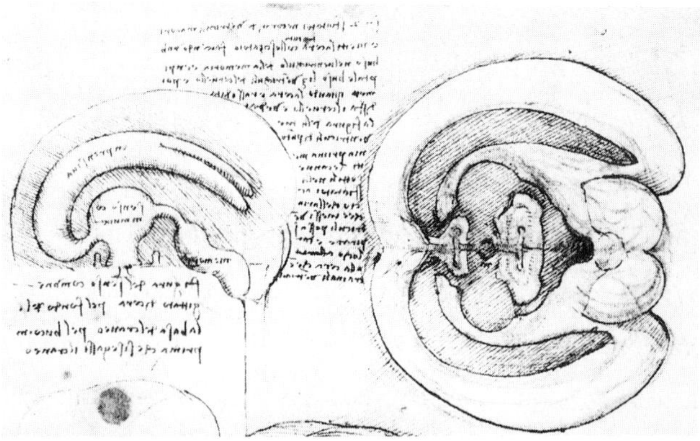

Fig. 1: Drawing by Leonardo da Vinci of wax cas of cerebral vent icles (Circa 149

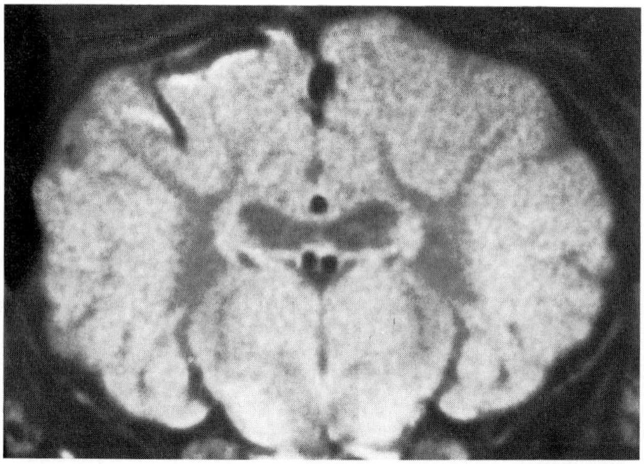

Fig.2: Magnetic resonance image of ox brain in coronal plane showing ventricles

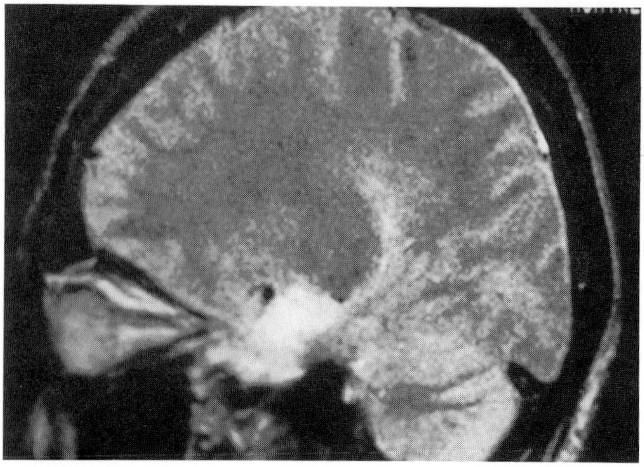

Fig.3: Case F: Subependymoma in amygdala on sagittal section

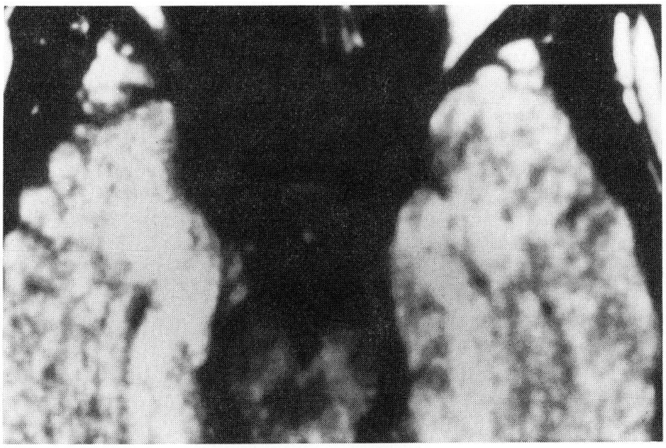

Fig.4: Case M: Ganglioglioma with high sigal intensity on axial view of temporal lobes

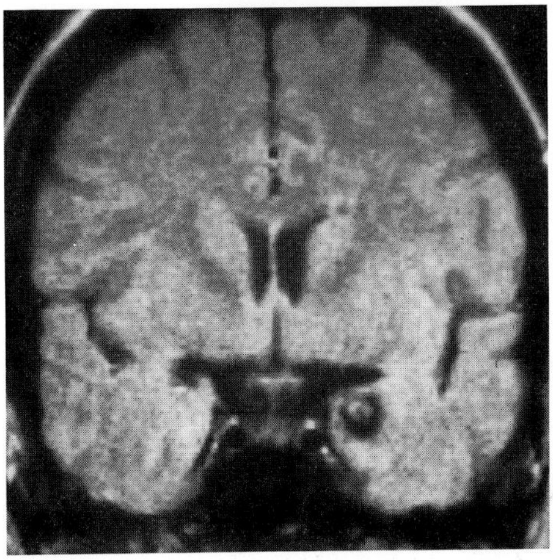

Fig.5: Case Q: Venous angioma in amygdala

larger series. The MRI results thus strengthen earlier views that seizures with automatism and amnesia are associated with small tumors, angiomas, or focal gliosis in the amygdala and antero-mesial temporal region. (4)

PHOSPHORUS-13 MAGNETIC RESONANCE SPECTROSCOPY (PMRS) OF BRAIN TUMORS

Phosphorus-13 is an atom that responds to the effect of a magnetic field, resembling in this respect the hydrogen atom. At high magnetic fields (1.5 tesla) the magnetic resonance imaging unit* can be rapidly adapted from an imaging mode to a spectroscopy operation. The PMRS spectrum consists of three ATP energy peaks and the phosphate derivatives, the highest of which represent phosphocreatine along with several phosphoesters and inorganic phosphate. The three peaks on the right are mainly related to cell energy and those on the left to cell membrane metabolism. The ATP peaks reflect the viability of the cell and undergo change only when the cell is severely damaged. A transient ischemia can temporarily effect the other peaks and restoration of circulation can reverse this effect (5,6).

MRS offers the important application of monitoring gliomas during chemotherapy with BCNU infused selectively into the carotid artery above the ophthalmic branch (7).

In ten intracranial tumors of different type proven histologically, MRS showed distinguishing features that promise to be of value in differential diagnosis (8). Meningiomas showed a low inorganic phosphate peak while pituitary adenoma (Fig.7) has in addition a high monoester compared to the normal spectrum (Fig.6). Low grade astrocytomas tend to have normal or slightly acid pH as compared to the alkaline pH of malignant gliomas. In a patient who had a pituitary adenoma removed some 20 years earlier, a recurrent mass appeared on MRI as a large adenoma. On MRS, however, the spectral configuration was indicative of a meningioma, with a low PME and PCR and a fairly low Pi; this was confirmed pathologically.

* Philips GYROSCAN, prototype research model supplied to the Montreal Neurological Institute and Hospital.

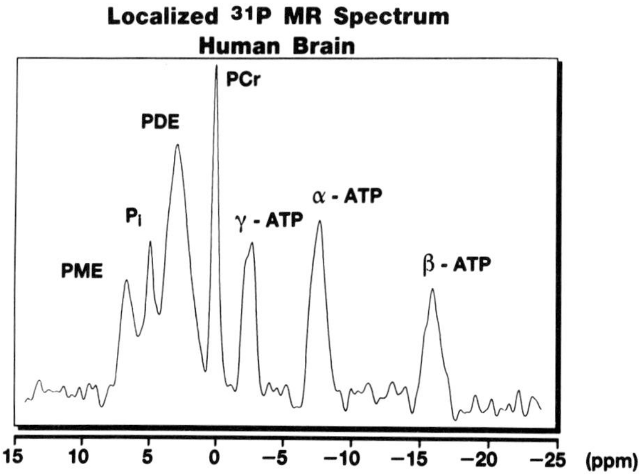

Fig.6: PMRS of normal brain, showing range of phosphorus containing compounds.See text

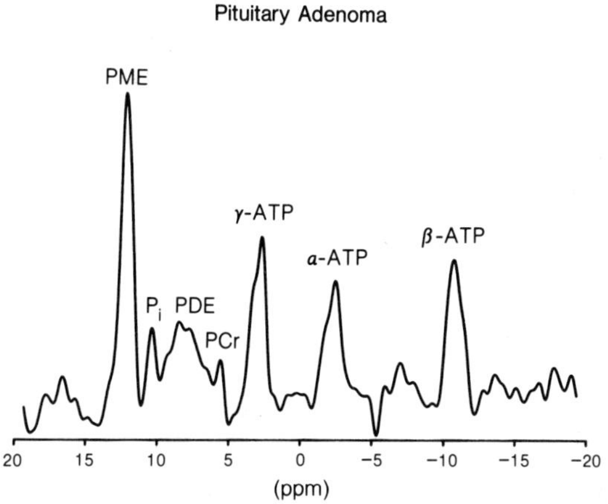

Fig.7: PMRS of Adenoma showing high PME, low PLr and PDE

CONCLUSIONS

1. Magnetic resonance imaging is capable of identifying small cerebral lesions including gliomas. Examples are given of lesions relating to focal cerebral seizures arising from the anterior and mesial temporal lobe; MRI detects 30% more of such lesions than CT.

2. MRI has provided also a new finding in temporal lobe epilepsy based on a signal intensity increase in the mesial temporal region

associated with astrocytic gliosis of the amygdala.

3. Phosphorus-31 magnetic resonance spectroscopy is a convenient non-invasive method for obtaining specific metabolic information including intracellular pH, on gliomas and other intracranial tumors which on preliminary studies promises to be useful in differential diagnosis.

ACKNOWLEDGEMENTS

This work was supported by a grant from the Department of Health and Human Services, USA, Grant No. NS.22230-03.

REFERENCES

1. Edelman R (1984) The Clinicians Guide to the Theory and Practice of NMR Scanning. Discussions in Neurosciences. 1(1): FESN Geneva.

2. Kuzniecky R, de la Sayette V, Ethier R, Melanson D, Andermann F, Berkovic S, Robitaille Y, Olivier A, Peters T and Feindel W (1987) Ann Neurol 22:341-347.

3. Mathieson G (1975) Advances in Neurol 11:163-185.

4. Feindel W, Robitaille Y, Ethier R, Quesney F (1988) Ann Meeting, Amer Assoc Neurol Surgeons, April.

5. Gadian DC (1982) Nuclear Magnetic Resonance and its Application to Living Systems. Clarendon Press, Oxford.

6. Feindel W, Frackowiak RSJ, Gadian DG, Magistretti PL and Zalutsky MR (eds.) (1985) Brain Metabolism and Imaging. Discussions in Neurosciences 2: (2) FESN, Geneva.

7. Arnold D, Shoubridge EA, Feindel W, Villemure JG (1987) Can J Neurol Soc 14:570-575.

8. Shoubridge EA, Arnold DL, Emrich JF, Villemure JG and Feindel W (1988) 7th Ann Meeting, Soc Mag Res in Med, August.

POSITRON EMISSION TOMOGRAPHY IN CEREBRAL GLIOMA

D.G.T. THOMAS

Neuro-Oncology Section, Department of Neurological Surgery, The National Hospital, Queen Square, London WC1. U.K.

In collaboration with the MRC Cyclotron Unit, Hammersmith Hospital, London W12.

INTRODUCTION

Positron emission tomography (PET) uses the short-lived positron-emitting radioisotopes of elements such as oxygen, carbon and fluorine as tracers, not only to image, but also to quantitate metabolic activity in the body. The method is relatively non-invasive and has been applied to measure tissue physiology and pathophysiology in man. Several studies have been performed in cerebral glioma by the author and collaborators at the MRC Cyclotron Unit at The Hammersmith Hospital in London as well as by other groups. The use of the method will be illustrated by examples of the information, some of it unexpected, which it has revealed about the physiology and metabolism of these tumours.

PET METHODS

The basis of PET is the administration of a positron-emitting radioisotope, usually as a biological tracer, and its subsequent detection and quantitation with an imaging device the positron computed tomograph. Most of the radioisotopes of those elements which are of biological interest, have short radioactive half-lives and it is therefore generally necessary to have on site a cyclotron for preparation of the tracers. Molecular oxygen is of great physiological importance and its isotope oxygen-15 (^{15}O) is the longest lived positron-emitting isotope of this element with a half-life of 2.1 minutes. ^{11}C and ^{13}N with half-lives respectively of 20.1 minutes and 10 minutes are the longest lived positron-emitting isotopes of 2 other elements of cardinal importance in the body's biochemistry. Although hydrogen does not have a similar isotope ^{18}F, with a half-life of 109 minutes, may be used as a suitable substitute for it in many tracer compounds. Some other isotopes of practical application may be produced without a cyclotron, for example ^{82}Rb with a half-life of 1.3 minutes may be produced from a generator containing ^{82}Sr, which itself has a 25 day half-life.

The positrons released by decay of such isotopes travel only very short distances before meeting negatively charged electrons. When this occurs both particles are annihilated and there is emission of 2 photons which proceed at an angle of 180° to each other. The positron computed tomograph is based on coincidence detection of these paired photons by photo multipliers in the scanning apparatus. The absorption characteristics of the tissue under examination may be calibrated by transmission computerised axial tomography in the same instrument. This allows computation of the activity which has been present in the biological tracer at a specific localisation and therefore produces quantitative data as well as a tomographic image. Although there are certain technical limitations on the spatial resolution of the method there has been considerable developments in technology so that now resolution of 2-3 mm is possible, whereas with earlier instruments typical resolution was 14-16 mm. The main factors which determine the special resolution of the tomograph are the size of the object of interest, the statistics of the observed physical data, and other aspects of the tomograph, like the response characteristics of the detectors and methods of attenuation correction. Further constraints are imposed on the interpretation of PET scans by the behaviour of the biological tracer in the body. It is desirable that the physiology and biochemistry of the tracer is understood so that mathematical models for its distribution in the body may be designed. For relatively simple physiological substances, for example oxygen and glucose, satisfactory tracer models have been established and quantitative data can be calculated. Where no valid tracer models have yet been agreed, for example ^{18}F-deoxyuridine ($^{18}FDUR$) only qualitative or semi-quantitative results may be obtained. In most valid models it is necessary also to obtain blood levels of the radioactive tracer and of the competing cold normal substrate, and this is usually achieved by estimations performed on serial arterial blood samples. The regional tissue isotope concentration of the tracer is obtained from the tomographic measurement involved in the particular model is computed using these pieces of data.

CEREBRAL BLOOD FLOW AND OXYGEN UTILISATION

The blood volume corrected, ^{15}O steady state inhalation technique involves the patient inhaling sequentially $C^{15}O_2$, $^{15}O_2$ and ^{11}CO, so that computation of regional blood flow (rCBF) and oxygen utilisation ($rCMRO_2$) together with the regional oxygen extraction fraction (rOER) and the regional blood volume (rCBV) may be determined (1, 2, 3).

In normal brain it is found that the rCBF and the $rCMRO_2$ is higher in cortex than in white matter. The rOER is uniform at about 40%. In normal brain there appears to be close coupling of oxygen supply and demand. Howver, in cerebral glioma it is found that there is a wide range in rCBF. The $rCMRO_2$ is generally low and the rOER is uniformly depressed below levels found in normal brain (4). Thus, in cerebral glioma oxygen demand is low and is not coupled to blood flow, although the PET method is macroscopic the implication is that these tumours, at least at the resolution of the method, have no evidence of hypoxia. A clinical implication of this finding is that hypoxic radiation sensitisers would probably produce no significant change in the tumours metabolism.

Gliomas may also affect the blood flow in remote brain areas. In the cortex of the contralateral hemisphere away from the tumour and depression of both rCBF and $rCMRO_2$ has been found when compared with normal control levels (2). The effect of administration of dexamethasone to patients with glioma has been to cause a slight further depression of rCBF and rCBV in the contralateral brain cortex, with corresponding increase in rOER. By contrast surgical decompression of the tumour at craniotomy has been shown to partially reverse the depression of rCBF and $rCMRO_2$. In patients receiving adjuvant radiotherapy following surgery for cerebral glioma an increase in rCBF in normal cortex with a constant $rCMRO_2$ has been found in the first few weeks. However, by 3 months there is a fall in rCBF and in rCBV with unaltered $rCMRO_2$. However, there is no indication of significant ischaemia in irradiated normal brain in these cases. In the glioma itself there tends to be a progressive fall in rCBF, rOER, $rCMRO_2$ and rCBV following radiation. These changes may reflect tumour cell death.

GLUCOSE METABOLISM

The methods for determining regional cerebral glucose utilisation (rCMRGlu) in humans employs ^{18}F-deoxyglucose (^{18}FDG) and the mathematical model allows expression of rCMRglu in quantitative units (5, 6, 7, 8). There remains controversy about the values of the lumped constant in this model as well as about individual rate constants for transport, phosphorylation and dephosphorylation. However, inspite of discussion about the exact quantitation of rCMRGlu, from PET studies in glioma a number of interesting findings have been made.

Di Chiro and others (9) found a direct correlation between glucose utilisation and grade of glioma. Thus, peak rCMRGlu found in Grade III and Grade IV gliomas was significantly higher than in Grades I and II. The same group identified areas of radionecrosis in brain, where rCMRGlu was low, in contrast to areas of recurrent tumour where rCMRGlu was high (10). The author and colleagues at the MRC Cyclotron Unit studied oxidative metabolism in cerebral glioma by using the ^{18}FDG method sequentially after determining oxygen metabolism by the ^{15}O method described above (11). The mean rCMRGlu in cerebral glioma was found to be similar to that in normal cortex with regional glucose extraction ratios (rGER) at normal levels. This contrasts with the depressed rOER described in the section above. In normal brain glucose and oxygen consumption is closely coupled. However, it appears in gliomas this is not the case and that there is an increase in nonoxidative metabolism of glucose. This probably represents glucose utilisation by glycolysis inspite of the presence of aerobic conditions within the brain. This striking difference in metabolic behaviour in the tumour compared with the normal surrounding tissue may possibly be of use in planning on a rational basis drug treatment using selective metabolic blockade.

BLOOD BRAIN BARRIER

Under normal circumstances the blood brain barrier is highly impermeable to potassium irons, and their passage is controlled by active membrane transport systems. ^{82}Rb is very similar in its permeability to potassium and may be traced by PET (12). In studies of cerebral glioma using this tracer it has been found that in cases where the blood brain barrier has broken down sufficiently to allow enhancement on CT scanning, following intravenous iodine contrast, that the PET ^{82}Rb scan will also show breakdown of the barrier. In those cases where there is no enhancement on CT there is also a relatively maintained barrier to ^{82}Rb. Yamamoto and colleagues (13) used Ga labelled EDTA to delineate cerebral tumours by PET. Several studies have been made of amino acid distribution in cerebral glioma using PET (14, 15, 16). It is likely that the distribution of these tracers reflects specifically increased amino acid uptake by active transport mechanisms at the capillary glial interface. Albumin uptake has been studied using PET by the author and collaborators and within the period of study over 2 hours no increased uptake of this large molecule was found. However, it is likely that over a longer time period some diffusion of the material into cerebral glioma might occur.

OTHER STUDIES

Some chemotherapy drugs, including BCNU, have been labelled and localised in brain tumour by PET (17, 18). Recently ^{18}F deoxyuridine has been applied as a tracer in PET to delineate tumours (19). This material is a precursor in DNA synthesis and uptake into tumour cells may follow specific metabolic pathways. The further application of this promising tracer awaits future development.

CONCLUSION

PET offers a method of studying the metabolism of human gliomas in situ relatively non-invasively. The pathophysiology, which it has revealed, may have interesting implications for design of therapy by surgery, by radiation and by drug treatment. The method is costly in resources and requires considerable patient compliance. It is unlikely to be of routine clinical application in the near future, although it is conceivable that technical refinements will make it more readily applicable for general use. If the information which it provides is shown to be of major use in planning individual patient management, there will be pressure to make PET more widely available.

REFERENCES

1. Frackowiak, R.S.J., Lenzi, G.-L. and Jones, T., et al. (1980) Quantitative measurements of regional cerebral blood flow and oxygen metabolism inman using ^{15}O and positron emission tomography: Theory, procedure and normal values. J. Comput. Assist. Tomogr. **4**:(6)727-736.

2. Lammertsma, A.A., Wise, R.J.S. Heather, J.D. et al. (1983) Correction of the presence of intravascular oxygen-15 in the steady state technique for measuring regional oxygen extraction ratio in the brain. 2. Results innormalsubjects and brain tumour and stroke patients. Cereb. Blood Flow Metabol. **3**:425-431.

3. Phelps, M.E., Huang, S.-C, Hoffman, E.J., et al. (1979) Validation of tomographic measurement of cerebral blood volume with C-11 labelled carboxyhemoglobin. J. Nucl. Med. **20**:328-334.

4. Ito, M., Lammertsma, A.A., Wise, R.J.S. Wise, et al. (1982) Measurement of regional cerebral blood flow and oxygen utilisation in patients with cerebral tumours using ^{15}O positron emission tomography: Analytical techniques and preliminary results. Neuroradiol. **23**:63-74.

5. Brooks, R.A. (1982) Alternative formula for glucose utilisation using labelled deoxyglucose. J. Nucl. Med. **23**:538-539.

6. Huang, S.-C., Hoffman, E.J., Phelps, M.E., et al. (1979) Noninvasive determination of local cerebral metabolic rate of glucose in man. Am. J. Physiol. **238**:69-82.

7. Phelps, M.E., Huang, S.-C., Hoffman, E.J., et al. (1979) Tomographic measurement oflocal cerebral glucose metabolic rate in humans with (F-18) 2-fluoro-2-deoxy-D-glucose: Validation of method. Ann. Neurol. **6**:371-388.

8. Reivich, M., Kuhl, D.E., Wolf, A., et al. (1979) The ^{18}F-fluorodeoxyglucose method for the measurement of local cerebral glucose utlisation in man. Circ. Res. **44**:17-137.

9. Di Chiro, G., De La Paz, R.L., Brooks, R.A., et al. (1982) Glucose utilisation of cerebral gliomas measured by (^{18}F) fluorodeoxyglucose and positron emission tomography. Neurol. **32**:1323-1329.

10. Patronas, N.J., Di Chiro, G., Brooks, R.A., et al. (1982) Work in progress: (^{18}F) Fluorodeoxyglucose and Positron Emission Tomography in the evaluation of radiation necrosis of the brain. Radiol. **144**:885-889.

11. Rhodes, C.G., Wise, R.J.S., Gibbs, J.M., et al. (1983) In vivo disturbance of the oxidative metabolism of glucose in human cerebral gliomas. Ann. Neurol. **14**:614-626.

12. Yen, C.K., Yano, Y., Budinger, T.F., et al. (1982) Brain tumour evaluation using Rb-82 and positron emission tomography. J. Nucl. Med. **23**:532-537.

13. Yamamoto, Y.L., Thompson, C.J., Meyer, E., et al. (1977) Dynamic positron emission tomography for study of cerebral haemodynamics in cross section of thc head using positron emitting ^{68}Ga-EDTA and ^{77}K. J. Comput. Assist. Tomogr. **1**:43-56.

14. Bustany, P., Chatel, M., Derlon, J.M., et al. (1986) Brain tumour protein synthesis and histological grades: A study by positron emission tomography with C-11-L-methionine. J. Neuro-Oncol. **3**:397-404.

15. Bergstrom, M., Collins, P., Ehrin, E., et al. (1983) Discrepancies in brain tumour extent as shown by computerised tomography and positron emission tomography using [^{68}Ga]-EDTA, [^{11}C]glucose, and [^{11}C]methionine. J. Comput. Assist. Tomogr. **7**:1062-1066.

16. Hubner, K.F., King, P., Gibbs, W.D., et al. (1981) Clinical investigations with carbon-11 labelled amino-acids using positron emission tomography in patients with neoplastic diseases. Med. Radionuclide Imaging IAEA (Vienna), 515-529.

17. Maeda, T., Kono, A., Kojima, M. (1972) Tumour scanning with ^{57}Co-Bleomycin. Radioisotopes **21**: 436-438.

18. Tyler, J.L., Yamamoto, Y.L., Diksic, M., et al. (1986) Pharmacokinetics of super selective intra-arterial and intravenous [^{11}C] BCNU evaluated by PET. J. Nucl. Med. **27**:775-780.

19. Gill, S. (1988) Personal communication.

© 1989 Elsevier Science Publishers B.V. (Biomedical Division)
Cerebral gliomas. G. Broggi and M.A. Gerosa, editors

THE MAIN MORPHOLOGICAL ASPECTS IN THE EVALUATION OF THE BRAIN TUMOR BIOPSIES

A. ALLEGRANZA, G. BROGGI, A. FRANZINI

Istituto Neurologico "C. Besta", Via Celoria 11, 20133 Milano (Italy)

INTRODUCTION

The use of biopsy in the diagnosis of neoplasias and tumor-like cerebral expanses has now become a daily practice in numerous neurosurgical centres.

As a serial stereotactic biopsy, it is considered essential as a pre-operative procedure in the case of deep seated cerebral tumoral lesions surgically not accessible because they are near vital structures, such as glial tumors of the basal nuclei, of the thalamus, in the vicinity of the third ventricle and of the brain stem.

Also during open surgical operations the removal of a bioptic sample may be sometimes necessary for diagnostic purposes, both because the neuroradiological data leave doubts as to the precise nature of the lesion and to ascertain the boundaries between normal and neoplastic tissue.

In stereotactic biopsies more than one sample is almost always taken to have a map of the various aspects the glial neoplasia can present. When, however, during an open surgical operation it is necessary to ascertain the histological type of the lesion and/or its boundaries with the healthy tissue, the number of samples varies depending on how the operating area appears to the neurosurgeon. The samples are processed following a pre-determined programme: extemporaneous cyto-histological wet smears can be made on a small part of each sample and examined at the microscope in a room adjacent to the operating room. The bulk of the sample is then dipped in a fixative liquid, or divided into more than one fixative depending on the purposes, the main one being to obtain permanent preparations on which to make a histological diagnosis using routine stainings and, if necessary, immuno-histochemical methods.

The cryostat is used when the samples to be examined are sufficiently consistent; this more often occurs during open surgical operations.

ACCURACY OF STEREOTACTIC BRAIN TUMOR BIOPSIES

The accuracy of the histological diagnosis in stereotactic biopsies chiefly depends on the adequacy of samples and on the training of the histopatologist.

In wet smears the accuracy is about 85% (1, 2)

On permanent preparations, where you can count on the help of immunohistochemical methods, the accuracy varies from 92 to 98%, depending on the Authors.

The cases in which the clinical and neuroradiological data reveal a clear tumoral pathology but the histological examination reveals a negative result are called false negatives.

The cause is attributed to errors in sampling and the response is given as normal tissue or inconclusive material. Edner (3) declares 9% of cases, Kleihues (4) 2.5% of cases and Bosch (5) 6.4% of cases.

The cases in which there is a discrepancy between the first histological bioptic diagnosis and the subsequent diagnosis -the latter based on bioptic sample or taken during an open surgery or autopsy- these are called pitfalls.

Most of the times the error is due to samples of tissue which are insufficiently representative of the lesion. Bosch (5) declared 2% of cases. Anyway the diagnostic error will be reduced to the minimum if the pathologist is acquainted with clinical data such as age, sex, location of the lesion and with neuroradiological data, as it is known that some neuroepithelial neoplasias have predilections for age and location, and in the case of metastasis both the age and the sex can be decisive in excluding such eventuality.

MORPHOLOGY

A - Permanent preparations

In the description of the morphology of bioptic samples in the case of gliomas (differentiated astrocytomas, anaplasic astrocytomas, glioblastomas, oligodendrogliomas and oligoastrocytomas, both differentiated and anaplasic) useful features for the differential diagnosis in the wider field of neuroepithelial neoplasias will be outlined.

The morphology of neuroepithelial tumors, and therefore also of gliomas in a strict sense, is often characterised by the way in which the neoplastic elements join each other and by the relationships which they can assume in their evolution with the pre-existing nervous tissue and with the meninges.

These aspects are distinguished by Scherer (6) in

1) primary structures, the way in which the "glioma" structures itself;

2) secondary structures, created by the "glioma" cells during its diffusion and which depend on the structure of the pre-existing nervous tissue; and in

3) tertiary structures linked to the interaction between the "glioma" and the mesenchymal tissue and the meninges.

1) The most typical primary or proper structures are the true rosettes of Bayley, called true so as to distinguish them from those which do not delimit a lumen or a similglandular cavity. Two types can be distinguished: Flexner's rosettes which characterise retinoblastomas (or neuroepitheliomas of the retina) and the rosettes of the ependymoma which are present in the epithelial variant of the ependymomas and in ependymoblastomas; these can be distinguished from the former if cilia and blepharoblasts are present.

Tubular-canalicular, similglandular structures are also present in the cerebral medullo-epitheliomas (or neuroepitheliomas) and in ependymoblastomas, sometimes in myxopapillary ependymomas of the cauda. The characterisation of these structures is based also in these tumours on whether blepharoblasts are present or not.

When the cells do not delimit a lumen, the structures are called pseudo-rosettes. Of these fundamentally two types can be distinguished: the Homer Wright's rosettes characteristic of neuroblastomas of the medullary of the suprarenal gland and of the thoracic-abdominal sympathetic ganglia. These structures may be also observed in the rare cases of neuroblastomas of the cerebral hemispheres and sometimes in medulloblastomas of the cerebellum. This finding also suggests the possible existence of supratentorial medulloblastomas.

The cells which form the Homer Wright pseudorosettes show a small hyperchromic nucleus and a slender cytoplasm which ends the centre of the structures in tangled fibrillated filaments which sometimes show affinity for silver impregnations. This suggests a possible neuroblastic differentiation of the elements.

The second type of pseudorosette is the perivascular type: the elements are arranged in a ring around small blood vessels to which they send a cytoplasmatic extension, which comes into contact with the vessel wall, whereas the nucleus is situated distally.

The perivascular pseudorosettes are characteristic of the cellular variant of the ependymomas and of the mixed cellular and epithelial variant. Besides, similar features can be found in the so-called true or primitive polar spongioblastomas of Russell and Cairns, which are rare and controversial neoplasias. Perivascular patterns can be found also in multiform glioblastomas and consequently in the case of bioptic microfragments they can represent elements of diagnostic uncertainty.

Among thc primary structures can be grouped some regressive processes of tumoral elements, such as the fibres of Rosenthal and the cytoid bodies, because they are characteristic of some neoplasias such as pilocytic astrocytomas especially of the cerebellum, subependymomas and the so-called gliomas of the optic nerve. Sometimes they can also be found in gangliogliomas.

Nevertheless, care is needed when it is a question of small bioptic fragments because the above-mentioned structures can be found in a peritumoral reactive gliosis, for example around craniopharyngiomas compressing the chiasma or the floor of the third ventricle, around teratomatous neoplasias of the pineal region, around cordomas of the clivus compressing the brainstem, in the vicinity of cerebellar and spinal hemangioblastomas, of spinal ependymomas, as well as around vascular dysplasia, cryptic angiomas, old cerebral infarction, foreign body granulomas, etc.

The morphology of the capillaries and the necrosis of the neoplastic tissue occupy a special place in the histological diagnosis of bioptic fragments.

The capillary vessels of low grade gliomas are slender, delicate and the endothelia appear normal. The presence of hypertrophic and hyperplastic endothelia is sign of angiogenesis and its proliferation can result in small coiled tangles of capillaries which remind the morphology of renal glomeruli. They are signs of anaplasia of the tumour and thus of malignity if the fragments under examination show the morphology of an astrocytoma. If small pieces of necrosis are observed in the same section the diagnosis of glioblastoma can be made.

Endothelial capillary proliferations can also be observed in oligodendrogliomas and in oligoastrocytomas, but in these cases, in order to formulate hypothesis of malignancy, it is necessary to find signs of cellular anaplasia and an evident mitotic index. The contemporary presence of necrosis will give the certainty that this is an anaplastic form. The endothelial proliferation of capillaries in the pilocytic astrocytomas of the cerebellum, however, does not signify a malignant evolution.

As regards the necrosis of tumoral tissue, this is essential in the diagnosis of bioptic microfragments. It is a necessary finding for the diagnosis of glioblastomas when faced with fragments indicating an astrocytoma. Necrosis in the glioblastomas may be widespread and therefore large necrotic strips can appear in the biopsies. The necrosis can also appear as a small perinecrotic pseudorosette and, if accompanied by other cytologic features -possibly confirmed by immunohistochemical methods present in other samples of the same case- it is decisive for the above-mentioned diagnosis. However, it must be careful because necrosis can be also observed in oligodendrogliomas and oligo-astrocytomas in malignant evolution, in malignant ependymomas, in medullo-blastomas and other primitive, malignant neuroepithelial tumours.

The differentiation from necrosis of metastasis of carcinomas, if also the clinical data are not in favour of such an eventuality, the age of patient, for example, is based on the morphology of the elements in necrosis or in necro-biosis: in the case of metastasis they almost always appear as cellular epitheliomorphous shadows.

2) The secondary structures differ depending on the structure of the nervous tissue which the "glioma" meets in its process of diffusion. Their analysis and their exact evaluation are particularly useful in cerebral biopsies, especially serial stereotactic biopsies, because they allow the estimation of the extension and progression of the process towards the deep tissues and the cortex.

The principal secondary structures are: perineuronal, perivascular, subpial, subependymal, peri- and intra-fascicular growth. The first three are found, in particular, in oligodendrogliomas and indicate a progression of the neoplasia in the cortex. Peri- and intra-fascicular growth occurs in the white matter and is well evidenced by staining methods for myeline sheaths as well as for the cells. As isolated findings, in the case of serial stereobiopsies, these types of growth do not allow a precise grading except when the cellular elements reveal clear signs of anaplasia together with an evident mitotic index: most of the times this means that the tumour has not been completely explored by the bioptic needle, or that it refers to the periphery of the lesion or the transition zone not yet clearly infiltrated.

3) The tertiary structures are represented by mesenchymal-vascular

proliferations in the necrotic or necrotic-haemorrhagic zones of glioblastomas and by cicatricial organisation of the necrosis -which sometimes can simulate similcarcinomatous aspects. This latter is a rare eventuality which should however be taken into consideration in the case of small bioptic fragments.

Similsarcomatous aspect can be assimilated to the tertiary structures. This sometimes happens when the gliomatous tissue invades the leptomeninges and simulates the proliferation of connective elements. This picture, however, is to be distinguished from the sarcomatous transformation of the connective-vascular stroma of the gliosarcomas. In such cases immunohistochemical stainings can be helpful, such as the immunoperoxidase GFAP, together with the demonstration of the VIII RAG factor.

B - Cytological and cytohistological preparations

There are greater diagnostic difficulties as regards wet smears and only daily training will allow satisfactory accuracy in diagnosis.

The principal diagnostic criteria concern:

a) cellular and nucleo-nucleolar polymorphism, the staining affinity of the chromatin, the presence and the number of typical and/or atypical mitoses, etc.;
b) the cellularity of the smear;
c) the cohesion of the elements and presence of the above-mentioned primary or proper structures of Scherer (true rosettes or pseudorosettes). These structures, however, are not often observed because the technique in preparing the smear tends in most cases to break them up;
d) the presence of Rosenthal fibres and cytoid bodies which, as stated previously, are characteristic of low-grade astrocytomas of the cerebellum. However these can be misleading because they are sometimes found in glial perilesional reactions;
e) the morphology of the blood vessels and their endothelia. Signs of angiogenesis indicate anaplasia of the process;
f) the presence of necrotic tissue indicates glioblastomas, but it is to be distinguished from necrosis of the metastatic processes;
g) the observation of one or more cellular populations. This data is important in the case of biopsies in the pineal region, a frequent location of germinomas. If the sample is adequate the diagnosis of this neoplasia is

easy even on wet smears, when large elements with vesicular nucleus and prominent nucleoli are mixed with lymphoid elements. The permanent preparation will confirm the diagnosis both on the basis of routine methods and the demonstration of PAS-positive granules in the large elements. Immunohistochemistry shows that in about 45% of the germinomas there is a positivity for the human chorionic gonadotropin (HCG) whilst the alpha-feto-protein (AFP) and the carcinoembryonic antigen (CEA) are negative (7).

However, the presence of more than one cellular population in lesion of this region does not always indicate a germinoma. In a wet smear of a case of sarcoidosis of the pineal region quite different elements were observed -some of which were clearly recognised as lymphocytes and others as large elements with a clear nucleus and nucleolus or with many nuclei sometimes arranged on periphery-. Only permanent preparations allowed a correct diagnosis (8).

Finally, a very difficult tumoral pathology to diagnose precisely through biopsy, both on fresh and on permanent preparations is that of the supra-tentorial tumours in children. Most of the times, particularly on smears, in the case of lesions situated in the cerebral hemispheres, the bioptic samples permit only to distinguish low-degree or malignant neuroepithelial neoplasias. In this second situation, if primary structures which refer to a determined form are absent, one must resort to the generic definition of primitive neuro-epithelial tumour (PNT) malignant, a term introduced by Hart and Earle in 1973 (9).

ACKNOWLEDGEMENTS

This study has been partially supported by the Consiglio Nazionale delle Ricerche, grant n. 87.01505.44 and by the Associazione "Paolo Zorzi" for Neurosciences.

REFERENCES

1. Adams JH, Graham DI, Doyle D (1980) Brain biopsy. The smear technique for the neurosurgical biopsies. Chapman and Hall, London
2. Allegranza A,Broggi G, Franzini A (1987) Tumori cerebrali. Atlante di citoistopatologia. Edi-Ermes, Milano
3. Edner G (1981) Stereotactic biopsy of intracranial space occupying lesions. Acta Neurochir 57: 213-234

4. Kleihues P, Volk B, Anagnostopoulos J, Kiesslin M (1984) Morphologic evaluation of stereotactic brain tumors biopsies. Acta Neurochir Suppl 33:171-181
5. Bosch DA (1986) Stereotactic techniques in clinical neurosurgery. Springer Verlag, Wien-New York
6. Scherer HY (1938) Structural development in gliomas. Am J Cancer 34: 333-351
7. Bjornsson J, Scheithauer BW, Okazaki H, Leech RW (1985) Intracranial germ cell tumors: pathological and immunohistochemical aspects of 70 cases. J Neuropathol Exp Neurol 44:32-46
8. Allegranza A, Mariani C, Grazioli L, Franzini A, Broggi G (1982) Accuracy of the wet smear technique in 110 stereotactic biopsies of intracranial tumors. Acta Neurologica (Napoli) 37:282-
9. Hart MN, Earle KM (1973) Primitive neuroectodermal tumors of the brain in children. Cancer 32:890-897

IMMUNOHISTOCHEMISTRY OF GLIAL TUMORS

ALESSANDRO MAURO, ALESSANDRO BULFONE

2nd Department of Neurology, University of Turin, Turin, Italy

INTRODUCTION

The goal of immunohistochemical techniques is to localize antigens in tissue reactions by means of specific polyclonal antisera or monoclonal antibodies (MAbs). The important technical advances in this field that have occurred in the last 10-12 years and the industrial production of a large amount of specific "immunological reagents" have greatly contributed to the success of immunohistochemistry. These techniques have became established as very valuable methods for increasing our ability to better interpret pathogical processes, especially in tumor histopathological diagnosis. The application of immunohistochemistry to neurooncology has been the subject of several recent reviews (2,6,25): the purpose of the present review, largely based on the experience gained in our laboratory, is to outline some of the problems derived from the application of immunohistochemistry to histopathological diagnosis.

TECHNICAL PROBLEMS

The first limitation in the detection of an antigen in tissue section derives from tissue preparation procedures. The ability of an antigen to react specifically with an antiserum varies significantly with the method of fixation and embedding employed. Type of fixative, duration of fixation, pH of the fixative and type of embedding material are the most important factors influencing the staining (33). Fortunately, in the immunohistochemistry applied to neurooncology, routine fixatives, such as formaldehyde or carnoy, followed by paraffin embedding, give satisfactory results (25). In some cases, particular antigens do not tolerate fixation and embedding treatments (8) and therefore a general principle of immunohistochemistry is that the absence of staining does not exclude the presence of an antigen. Enzymatic predigestion of routily fixed tissue can, in some cases,restore the antigenicity producing results similar to those obtained with frozen sections (17).

An exhaustive discussion of the different immunohistochemical techniques is outside the scope of this review. Briefly, immunofluorescence,immunoenzymatic techniques and avidin-biotin methods have all been successful in the demostration of antigens in glial tumors. Nevertheless in our experience immunoenzymatic methods are superior to immunofluorescence in diagnostic immunohistochemistry because they produce simultaneously good specificity and sensitivity with better morphology and stable stainings. On the contrary immunofluorescence can be performed quickly and therefore is useful for rapid tests or experiments.

Obviously the central concern of immunohistochemistry is to obtain specific binding of antibodies to the expected antigen and to avoid non-specific binding. Therefore methods controls, designed to test whether non-specific bindings has occurred, are extremely important in immunohistochemistry. Immunoblotting is particularly useful to this purpose .

The abundance of antibodies commercially available complicates the choice between polyclonal antisera and MAbs. For diagnostic purpose, MAbs are not necessarily superior to carefully purified polyclonal antisera (2), for at least two reasons . Firstly, the unquestionable increased specificity of MAbs does not imply a comparable increase sensitivity; secondly the single epitope that specifically binds to the MAb may occur on similar but not identical proteins,giving rise to cross-reactions and thus to misinterpretations. For these reasons it must be emphasized that specificity and sensitivity of both polyclonal antisera and MAbs must be similarly tested before applying them to diagnostic pathology.

DIAGNOSTIC APPLICATIONS

During the last twenty years a large number of antisera and MAbs have been raised against antigens of the nervous system.Among these antigens,comprising structural and membrane proteins,enzymes,neurotrasmitters,extracellular matrix components,etc. only very few can be considered specific for a single cell type or structure of the nervous system.Despite this "relative specificity", that becomes even more problematic in pathological nervous tissue, the recognition of some of this antigens may be very helpful in diagnostic neurooncology. Generally these antigens are utilized as cytotypic "markers",demostrating similarities or differences between neoplastic cells and normal putative elements.

Glial Markers

S-PROTEIN . The so-called S-100 (or 14-3-2 protein) was the first protein to be considered specific for the nervous system . Its presence in astrocytes and oligodendrocytes, as well in ependimal and Schwann cells, is well documented, while there is no general agreement about its occurence in neurons. The studies reporting the distribution of S-100 protein in human brain tumors are relatively few (21,34). The distribution of S-100 protein in glial tumors is very similar to that of GFAP, but its diagnostic value is strongly limited by the finding that the protein can be present also in non glial tumor cells of both central and peripheral nervous system, as well as in tumor cells of non neuroepithelial origin, carcinomas included. From these observation it derives that the usefulness of S-100 protein in diagnostic neurooncology is questionable and probably limited to circumnscribed problems such as the differential diagnosis of anaplastic tumors of PNS (2).

Glial Fibrillary Acidic Protein . Glial fibrillary acidic protein, the classical subunit of the intermediate filaments of glial cells, is the most reliable and the most widely used antigen for clinical and experimental purposes in neurooncology. There is general agreement about the value of its immunohistochemical demostration for the resolution of diagnostic neurooncological problems (1,2,5, 9,21). In CNS antibodies recognizing GFAP stain normal astrocytes of white and grey matter, Bergman glia of the cerebellum and tanicytes. GFAP can be demostrated in tumors composed of cells of astroglial origin including astroblastomas, subependymal giant astrocytomas, subependymomas, pleomorphic xanthoastrocitomas (2). A positive reaction for GFAP is demonstrable in the astrocytic cells of mixed tumors,such as gliosarcomas,sarcogliomas,oligo-astricytomas,angiogliomas and gangliogliomas.The staining for GFAP in anaplastic astrocytomas and glioblastomas is, in general,extremely variable between different cases and between different areas in the same case. the presence of GFAP can considered an expression of the degree of differentiation of glial cells and therefore it is well understandable that cells of the less differentiated areas of glioblastomas do express the protein (5,9,26,35). A practical problem in the interpretation of the results of GFAP immunohistochemistry derives from the observation that in several istances GFAP is found in non astroglial neoplastic cells. The relatively frequent staining for GFAP in oligodendrogliomas is well known. In some cases it can be attributed to reactive astrocytes, or to the astrocytic component of an oligoastrocytoma, but in 40%-50% of oligodendrogliomas GFAP staining corresponds to true neoplastic oligodendrocytes (12). This finding is not surprising taking into account that normal human fetal oligodendrocytes express GFAP, transiently before myelinogenesis (3). Ependymomas generally contain a consistent number of GFAP immunoreactive cells, frequent in perivascular "pseudorosettes" with a characteristic pattern . Moreover a number of non glial tumors may contain GFAP positive cells: choroid plexus papillomas, hemangioblastomas, pinealoblastomas, medulloblastomas. The GFAP staining of some stromal cells of hemangioblastomas has been regarded has a demonstration of their glial origin (15), or conversely as a manifestation GFAP uptake by stomal cells of angiogenic origin (7). The GFAP-positive cells that can be seen frequently in medulloblastomas have been considered expression of glial differentiation (2) or entrapped subependimal glial cells (24) or only reactive astrocytes (24, 30). WE have observed a clear GFAP staining in the cells of small cartilagineous foci in a CNS teratoma (38). These findings, in our opinion,can be considered as "other specificities" of GFAP and in general do not represent a limitation to the helpfulness of GFAP demostration in diagnostic neurooncology.

Oligodendroglial Markers : CA-C, MAG, MBP, LEU 7

Three proteins have been considered to be possible specific markers for oligodendrocytes, namely Carbonic-Anhydrase C (CA-C) (16), myelin-assiociated glycoprotein (MAG) (33), and myelin basic protein (MBP) (14), but practically they seem useless in the diagnosis of human oligodendrogliomas for different reasons. The staining of human normal and neoplastic oligodendrocytes with CA-C antibodies is inconsistent, while endothelial and neoplastic cells in several nonglial tumors, as well as reactive astrocytes, are frequently CA-C positive (20).

Our experience in human brain tumors is in line with these results (unpublished data). Moreover in our laboratory it has been demonstrated that also experimental oligodendrogliomas induced in rat by ENU are not stained by CA-C antisera, in contrast with the specific staining of normal rat oligodendrocytes (11).Therefore CA-C cannot be used has a specific marker for human normal or tumoral oligodendrocytes .

During development in man MBP can be found only in immature oligodendrocytes, While in adult it si present only in myelin sheaths. In our experience neoplastic cells of human and experimental oligodendrogliomas are MBP negative (unpublished data).

Recently Motoi et. al. (19) suggested that the LEU 7 (or HNK-1) monoclonal antibody generated against a human T-cell line and cosidered amarker of human natural killer cells, colud be a valuable marker for oligodendrogliomas. Following studies demonstrated that LEU 7 antibody stains stongly oligodendrogliomas but reacts also with the most tipes of the other neurogenic tumors (22).

The obvious conclusion is that, until now,there are no specific "markers" for the diagnosis of oligodendrogliomas.

Non-Glial Markers

Several antigens, non glial specific, may be useful in the diagnosis of gliomas. Among putative neuron specific antigens, neurofilaments and neuron specific enolase received the most attention. Antibodies and antisera against the three subunits of neurofilaments are undoubtedly neuron specific but different antibodies show different patterns of staining between axons and perikaryon as well as between the various types of neurons (32). However the demonstration of neurofilaments (NFs) can be extremely helpful in the demonstration of ganglion cell component of gangliogliomas (23).

On the contrary, the role of neuron specific enolase (NSE) in thr diagnosis of neuroepithelial tumors must be questioned: while it is generally accepted that in normal nervous system NSE is a specific marker of neurons, this is not the case in pathological conditions. The enzyme usually can be demonstrated in a number of non neuronal cells occurring in CNS and extraneuronal tumors, as well as in reactive astrocytes (36,37).

Vimentin is the first intermediate filament protein to appear during embryological development, regardless of cell life. In immature astrocytes, vimentin is the major cytoskeletal component (4) and is later replaced by GFAP. Coexpression of vimentin and GFAP has been demonstrated in mature astrocytes (31), while normal mature ependyma contains only vimentin (31,27,28). Vimentin and GFAP are coexpressed in tumors of astroglial derivation and in ependymomas (28,13). In astrcytic gliomas the staining for vimentin occurs both in glial and in vascular elements. In glial cells the distribution of vimentin is very similar to that of GFAP, so that, in our opinion, vimentin cannot be considered a marker of dedifferentiation in gliomas, as previously supported (23).

Neovascularization and endothelial proliferation are common features of malignant gliomas and from the vasculature of these tumors the neoplastic mesodermal component of the gliosarcoma takes origin (29). At least three antigens have been successfully used for the identification of vessel walls in gliomas: Factor VIII Related Antigen (FVIII/RAg), Laminin and fibronectin.Antibodies against FVIII/RAg a specific marker of endothelial cells, stain normal and hyperplastic endothelial cells (17,18). In gliosarcomas the sarcomatous proliferation is FVIII/RAg negative but the staining seems to progressively decrease from endothelial proliferation of larger glomeruli to fibrosarcomatous areas (29). Laminin is a glycoproteic component of basement membranes. Its value in demonstrating vessels of normal CNS and of neuroepithelial tumors has been confirmed (10). In gliomas anti-laminin sera show vessel proliferation, characterized by multilayered basement membranes partly surrounding proliferating endothelial cells. In gliosarcomas a basement membrane with few discontinuities separate glial areas from mesodermic areas (29). Fibronectin is the major protein of the extracellular matrix and in normal CNS can be demonstrated in vessel walls, not only in basement membranes (17). In gliosarcomas fibronectin is diffusely present in sarcomatous ares that easily can be differentiated from gliomatous ones, fibronectin-negative and GFAP -positive (29).

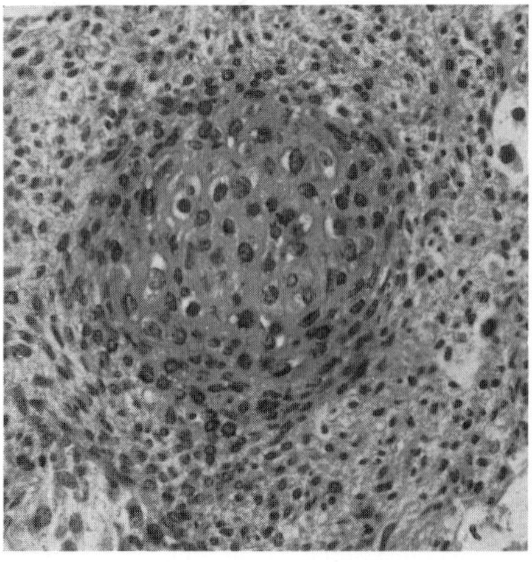

Fig. 1. GFAP: positive cells in cartilage of a teratoma (PAP-DAB)

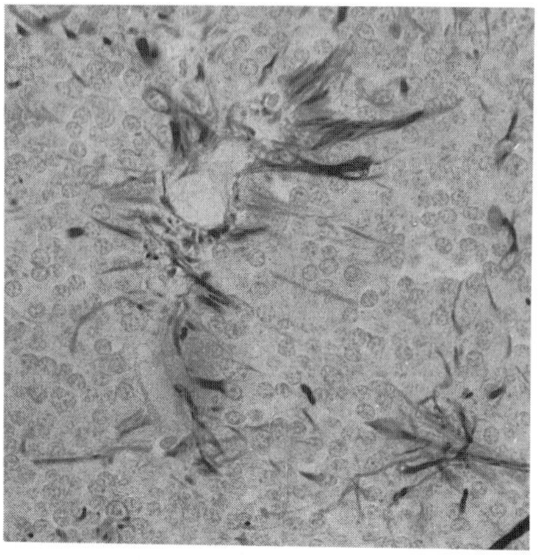

Fig. 2. GFAP: ependymoma with positive cells (PAP-DAB)

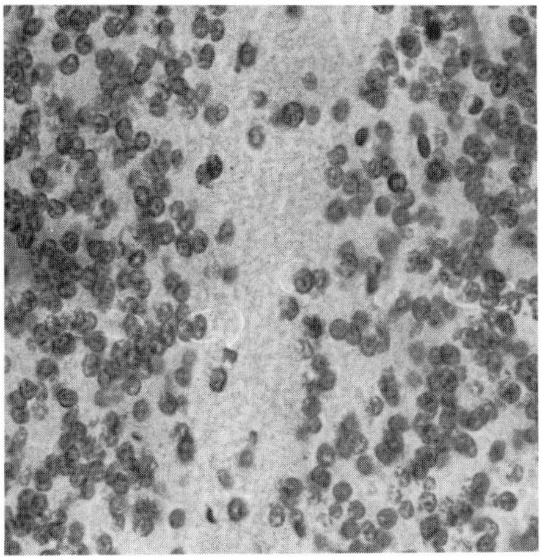

Fig. 3. CA-C: positive oligodendrocytes in human cerebellum

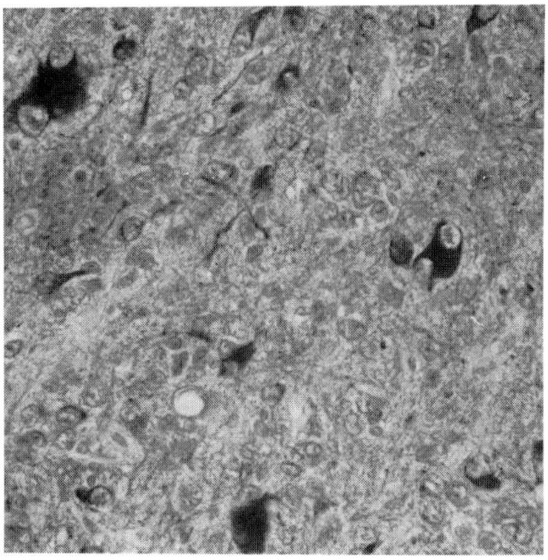

Fig. 4. NF: positive cells in a ganglioglioma

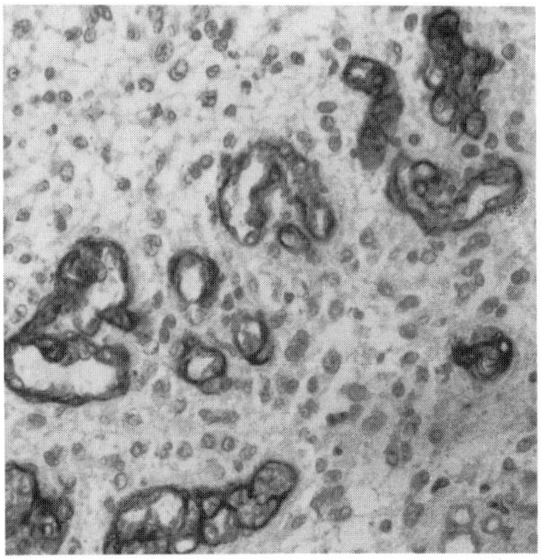

Fig. 5. FVIII/RAg: glioblastoma with positive cells of glomeruli

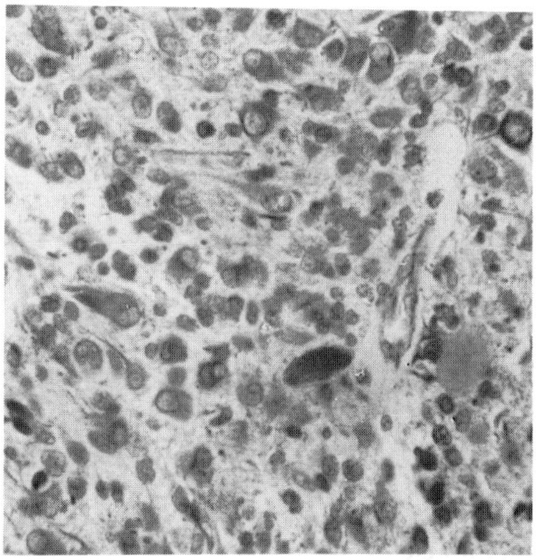

Fig. 6. Vimentin: positive cells in anaplastic astrocytoma

ACKNOWLEDGEMENTS

Supported in part by a grant of the Italian National Research Council (CNR), Special Project "Oncology", contract N° 87.01446.44, Rome, by the Italian Association for Cancer Research (AIRC)

REFERENCES

1. Bignami A, Schoene WC (1981) In: De Lellis RA (ed) Diagnostic Immunohistochemistry. Masson Publishing, USA, pp 213-225
2. Bonnin JM, Rubinstein LJ (1984) J Neurosurg 60:121-133
3. Choi BH, Kim RC (1984) Science 223:497-499
4. Dahl D, Rueger DC, Bignami A (1981) Eur J Cell Biol 24:191-196
5. De Armond SJ, Eng LF, Rubinstein LJ (1980) Pathol Res Pract 168:374-394
6. De Armond SJ, Eng LF (1984) Progr Exp Tumor Res 27:92-117
7. Deck JHN,Rubinstein LJ (1981) Acta Neuropathol 54:173-181
8. Dixon RG, Eng RF (1981) J Comp Neurol 201:15-24
9. Eng LF, Rubinstein LJ (1978) J Histochem Cytochem 26:513-522
10. Giordana MT, Germano I, Giaccone (1985) Acta Neurophatol 67:51-57
11. Giordana MT, Schiffer D, Mauro A, Migheli A (1986) In: Walker MD, Thomas DGT (ed) Biology of brain tumors. Martinus Nijhoff Pub, Boston, pp 121-129
12. Herpers MJ, Budka H (1984) Acta Neuropathol 64:265-272
13. Herpers MJ, Ramaekers FCS et al. (1986) Acta Neuropathol 70:333-339
14. Itoyama Y, Sternberger NH et al. (1980) Ann Neurol 7:157-166
15. Kepes JJ, Rengachary SS, Lee SH (1979) Acta Neuropathol 47:99-104
16. Kumpulainen T, Dahl D et al. (1983) J Histochem Cytochem 31:879-886
17. Mauro A et al. (1984) Histochemistry 80: 157-163
18. Mc Comb RD et al. (1982) J Neuropathol Exp Neurol 41:479-489
19. Motoi M, Yoshino et al. (1985) Acta Neuropathol 66:75-77
20. Nakagawa Y, Perentes E, Rubinstein LJ (1986) Acta Neuropathol 72:15-22
21. Nakamura Y, Becker LE, Marks A (1983) J Neuropathol Exp Neurol 43:136-145
22. Perentes E, Rubinstein LJ (1986) Acta Neuropathol 69:227-233
23. Roessmann U, Velasco ME et al (1983) J Neuropathol Exp Neurol 42:113-121
24. Schiffer D,Giordana MT, Mauro A et al (1983) Tumori 69:95-104
25. Schiffer D, Giordana MT, Mauro A et al (1986) Bas Appl Histochem 30:253-265
26. Schiffer D, Giordana MT et al (1986) Tumori 72;163-170

27. Schiffer D, Giordana MT et al (1986) Brain Res 374:110-118

28. Schiffer D, Giordana MT et al (1986) Acta Neuropathol 70:209-219

29. Schiffer D, Giordana MT et al (1984) Acta Neuropathol 63:108-116

30. Schindler E, Gullotta F (1983) Virchows Arch 398:263-275

31. Shaw G et al (1981) Eur J Cell Biol 26:68-82

32. Sternberger LA, Sternberger NH (1983) Proc Natl Accad Sci USA 80:6126-6130

33. Sternberger LA (1979) 2nd Wiley ed Immunohistochemistry New York

34. Takahashi K et al (1984) Virchows Arch 45:385-396)

35. Velasco ME, Dahl D et al (1980) Cancer 45:484-494

36. Vinores SA et al (1984) Arch Patol Lab Med

37. Vinores SA, Rubinstein LJ (1985) Neuropathol Appl Neurobiol 11:349-359

38. Mauro A, Germano I, Giaccone G (1985) XXI Meeting Ital Ass Neuropatol Siena pp123

Cerebral gliomas. G. Broggi and M.A. Gerosa, editors

PROGNOSTIC FACTORS IN CEREBRAL ASTROCYTIC GLIOMAS

R. SOFFIETTI, A. CHIO'

2nd Department of Neurology, University of Torino, Torino, Italy

INTRODUCTION

To study the prognostic factors means to study the influence of the clinical, radiological, biopathological and therapeutic factors on the course of a disease. In clinical oncology survival time is the major endpoint against which prognostic factors are assessed, but sometimes others endpoints, such as time until recurrence or progression or the probability of a response to therapy, are used. The variables must be examined not only individually for their influence on the outcome, but also simultaneously in a model of multivariate analysis in order to detect their relationships and their independent weight when acting together (1). The aim is that of forming, within a given tumor category, different "risk groups", defined by a combination of multiple prognostic factors. "Risk groups" (Prognostic subgroups), i.e. groups of patients with different probability of survival, could differently benefit from the same therapy.

PROGNOSTIC SIGNIFICANCE OF MALIGNANCY IN ASTROCYTIC GLIOMAS.

Histologic grading has been widely applied in the past to evaluate surgical biopsies of astrocytic gliomas (2-3), but its prognostic significance has not clearly emerged. Using Kernohan's grading system (2), most authors failed to demonstrate definite differences in survival between grades 3 and 4 (4-5-6-7). Recent studies, mostly from cooperative groups (RTOG-ECOG, BTSG) have demonstrated that the difference in survival is highly significant when malignant gliomas are subdivided in glioblastomas and anaplastic astrocytomas by the presence of necrosis, both in surgical (7-8) and stereotactic (9) biopsies. The difference in prognosis is maintained after reoperation for recurrence (10-11), but does not emerge is life expectancy is too short, as in patients with deep or midline tumors or in those with lobar tumors not adequately treated by radiotherapy (9). When malignant (anaplastic) features are absent in an astrocytic glioma, the term astrocytoma (well-differentiated) has been increasingly used instead of distinguishing a grade 1 and a grade 2 tumor (12-13-14-15-16). Most studies have failed to show any difference in survival between grade 1 and grade 2 (17-18-5-

19), though some have alternatively attributed a better prognosis to grade 1 (20-21-2) or grade 2 (22). In terms of prognosis the major problem is that of distinguishing the well-differentiated astrocytoma from the anaplastic one : it has been clearly demonstrated (23) that, glioblastomas apart, the presence of malignant features in an astrocytic glioma is a prognostically more important variable that clinical data (extent of surgery, performance status). Moreover, a slight and localized malignancy in biopsies seems to influence in the same way that an extensive malignancy does (23), and different features of malignancy (23) or degrees of anaplasia (24) do not differ significantly in affecting survival. The only factor, limiting unpredictably the prognostic significance of histologic diagnosis in astrocytic gliomas, is the morphologic variability from region to region, so that, in any individual case, one can never rule out the possibility of a sampling error unless the tumor is a glioblastoma.

PROGNOSTIC FACTORS IN WELL-DIFFERENTIATED ASTROCYTOMAS.

Few data are available on prognostic factors in well-differentiated astrocytomas of cerebral hemispheres (excluding the pilocytic subtype). We recently analyzed the prognostic importance of histologic and clinical factors in 85 of these cases (25). Cell density (low or medium), nuclear polymorphism (absent or slight), vessel frequency (normal or increased) and perivascular infiltrates (present or absent) did not affect survival. The presence of few mitoses ($<$5x10 high power fields) did not modify the prognosis, as already reported by Levy and Elvidge (26). Vessel size is in our experience the only histologic factor of prognostic significance both individually and after multivariate analysis. The inverse relationship of survival with the occurrence of vessels with variable size could indicate a sampling error at biopsy.

Clinical factors of prognostic significance after multivariate analysis in retrospective studies are young age, high performance status and gross total surgical removal (18-19-27-28-25). The efficacy of radiotherapy in incompletely resected tumors has been suggested in retrospective series lacking an extensive analysis of prognostic factors (29) and needs further investigations (prospectic and randomized). The prognostic importance of tumor enhancement on CT, originally denied by Silverman and Marks (30), has been reported by Piepmeier (28) with a statistical significance that remains positive after adjustment for age: astrocytoma enhancing on CT appears to carry a poor prognosis. Prognostic subgroups have been identified in large series of astrocytomas, with great variability in prognosis: 5-year survival ranging from 35% to 69% (19) and median survival from 383 to 1533 days (25). Law's analysis (19) has shown that radiotherapy, not useful when considering the whole population, was of clear benefit in a small subgroup, i.e. in older patients with incompletely removed tumors.

Malignant transformation or progression (toward the anaplastic astrocytoma or the glioblastoma) represents in the natural history of well-differentiated astrocytoma the true factor limiting the possibility of predicting survival by statistical methods. The exact risk of this phenomenon is unknown: the percentage of astrocytomas which showed anaplastic areas at the second biopsy or at autopsy varies in the different series from 13% to 85% (31-32-19-33-28-25). At this time there is no reliable method that can be used to predict this change in the tumor's biological activity (28).

PROGNOSTIC FACTORS IN ANAPLASTIC ASTROCYTOMAS

There is little information on prognostic factors in anaplastic astrocytomas separately from glioblastomas; in most series the two types have been analyzed together. Three studies are available on prognostic importance of histologic factors (8-34-35) and in all, the presence of vascular (endothelial) proliferations gives a poorer prognosis. Discrepancies (probably due to the low number of cases) still persist when the importance of the vascular proliferations is analyzed separately for the treatment received (surgery alone or surgery and radiotherapy.) Cell density, nuclear polymorphism and perivascular infiltrates seem devoid of any significance, whereas only in the experience of Fulling and Garcia (34) a high number of mitoses is associated with a shortening of survival.

Age and performance status are the clinical factors that generally emerge as prognostically important (8-36-11-35). Surgical removal (total or partial) gives better results that biopsy alone in irradiated patients (36), whereas adjuvant radiotherapy with doses >45 Gy improves significantly survival after surgical removal (35). The true value of the extent of surgery (total removal versus partial removal) in patients treated with high radiation doses (55-60 Gy) is still unclear. A complete versus a partial response to radio-chemotherapy on CT has been recently suggested as prognostically important (37).

PROGNOSTIC FACTORS IN GLIOBLASTOMAS

Glioblastomas containing microcysts, pseudopalisading, cells with astrocytic differentiation and large areas of better differentiated glioma show a longer survival than those homogeneously composed of small cells or those with small median nuclear size (38). The better prognosis associated with the occurrence of calcifications (39) has not been confirmed (38). The presence of giant cells, independently from their glial or sarcomatous nature, correlates positively with survival (40). An inversion of the prognostic value of cellularity has been observed by Burger and Vollmer (40) between glioblastomas treated with surgery alone and those treated with surgery, radio- and chemotherapy: this raises the more general problem of considering the impact of specific treatments, even if not curative, when evaluating prognostic factors. Perivascular lymphocytic infiltrates do not seem to correlate with survival in glioblastomas (40-38), whereas conflicting correlations have been the rule in the past when analyzing together Grade 3 and 4 tumors or glioblastomas and anaplastic astrocytomas (41-42-43-44-45). Conflicting data are available on the prognostic value of some kinetic parameter such as labeling index with tritiated thymidine: suggested as positive by Hoshino and Wilson (46), it has been recently denied by Bookwalter et al. (47). The existence of significant correlations between DNA pattern and survival remains to be defined (48-49-50).

Age, performance status and the extent of surgery seem to be the most important factors that predict survival in patients treated by multimodality therapy (1). Also in glioblastomas the response to radio- and chemotherapy on CT could be prognostic (37).

In the last years interesting data have been reported on the prognostic importance of CT and PET parameters in malignant gliomas considered as a whole. Tumor size on CT (both alter surgery and after radiation therapy) in the large series of BTSG (51) has emerged as an independent prognostic variable following age, performance status and pathology. In a PET study (52) the mean survival time of patients with tumors exhibiting high glucose utilization was 5 months, whereas patients showing lower glucose utilization had a mean survival period of 19 months. No data are yet available on the prognostic usefulness of magnetic resonance imaging.

ACKNOWLEDGEMENTS

Supported in part by a grant of the Italian National Research Council (CNR), Special Project "Oncology", Contract n° 87.01446.44, Roma, by the Italian Association for Cancer Research (AIRC) and by the CSI-Piemonte, Consorzio per il Sistema Informativo.

REFERENCES

1) Byar D, Green S, Strike T (1983) In: Walker MD (ed), Oncology of the Nervous System. Martinus Nijhoff, Boston, pp. 379-385

2) Svien HJ, Mabon RF, Kernohan JW (1949) Proc Staff Mayo Clinic 24:54

3) Ringertz N (1950) Acta Pathol Microbiol Scand 27:51

4) Weir B (1973) J Neurosurg 38:448

5) Scanlon PW, Taylor WF (1979) Neurosurgery 5:301

6) Salcman M 51982) Neurosurgery 10:454

7) Nelson JS, Tsukada Y, Schoenfeld D (1983) Cancer 52:550

8) Burger PC, Vogel FS, Green SB (1985) Cancer 56:1106

9) Coffey RJ, Dade Lunsford L, Taylor FH (1988) Neurosurgery 22:465

10) Ammirati M, Galicich JH, Arbit E, Liao J (1987) Neurosurgery 21:607

11) Harsh GR, Levin VA, Gutin PH, Seager M, Silver P, Wilson CB (1987) Neurosurgery 21:615

12) Schiffer D, Fabiani A (1975) I Tumori Cerebrali Il Pensiero Scientifico, Roma, pp. 140-242

13) Russell DS, Rubinstein LJ (1977) Pathology of Tumours of the Nervous System. Arnold, London, pp. 146-244

14) Zuelch KJ (1979) Histological Typing of Tumours of the Central Nervous System. WHO, Geneva pp. 43-45

15) Burger PC, Vogel FS (1982) Surgical Pathology of the Nervous System and its Coverings. John Wiley and Sons, New York, pp. 226-268

16) Fulling KH, Nelson JS (1984) Seminars in Diagnostic Pathology 1:152

17) Gol A (1961) J Neurosurg 18:501

18) Weir B, Grace M (1976) Canad J Neurol Sci 3:47

19) Laws ER, Taylor WF, Clifton MB, Okazaki H (1984) J Neurosurg 61:665

20) Leibel SA, Sheline GE, Wara WM, Boldrey EB, Nielsen SL (1975) Cancer 35:1551

21) Stage WS, Stein JJ (1976) AJR 120:7

22) Fazekas JT (1977) Int J Radiation Oncology Biol Phys 2:661

23) Schiffer D, Chiò A, Giordana MT, Leone M, Soffietti R (1988) Cancer 61:1386

24) Garcia DM, Fulling KH, Marks JE (1985) Cancer 55:919

25) Soffietti R, Chiò A, Giordana MT, Vasario E, Schiffer D (1988) Neurosurgery, in press

26) Levy LF, Eldvidge AR (1956) J Neurosurg 13:413

27) Cohadon F, Aouad N, Rougier A, Vital C, Rivel J, Dartigues JF (1985) J Neurooncol 3:105

28) Piepmeier JM (1987) J Neurosurg 67:177

29) Morantz RA (1987) Neurosurgery 20:975

30) Silverman C, Marks GE (1981) Radiology 139:211

31) Marsa GW, Goffinet DR, Rubinstein LJ, Bagshaw MA (1975) Cancer 36:1681

32) Mueller W, Afra D, Schroeder R (1977) Acta Neurochir (Wien) 37:75

33) Loftus CM, Copeland BR, Carmel PW (1985) Neurosurgery 17:19

34) Fulling KH, Garcia DM (1985) Cancer 55:928

35) Soffietti R, Chiò A, Giordana MT, Vasario E, Schiffer D (1988) Clin Neuropathol, in press

36) Nelson DF, Nelson JS, Davis DR, Chang CH, Griffin TW, Pajak TF (1985) J Neurooncol 3:99

37) Murovich J, Turowski K, Wilson CB, Hoshino T, Levin V (1986) J Neurosurg 65:799

38) Burger PC, Green SB (1987) Cancer 59:1617

39) Takeuchi K, Hoshino K (1977) Acta Neurochir (Wien) 37:57

40) Burger PC, Vollmer RT (1980) Cancer 46:1179

41) Maunoury R, Vedrenne C, Constans JP (1975) Neurochirurgie 21:213

42) Guidetti B, Palma L, Di Lorenzo N (1977) in Carrea R, Le Vay D (eds) Proceedings of the Sixth International Congress of Neurological Surgery, Sao Paulo

43) Schiffer D, Cavicchioli D, Giordana MT, Palmucci L, Piazza A (1979) 65:119

44) Brooks WH, Maresbery WR, Gupta GD (1978) Ann Neurol 4:219

45) Safdari H, Hochberg FH, Richardson EP (1985) Surg Neurol 23:221

46) Hoshino T, Wilson CB (1979) Cancer 44:956

47) Bookwalter JW, Selker RD, Schiffer L, Randall, Iannuzzi D, Kristofik M (1986) J Neurosurg 65:795

48) Frederiksen P, Reske-Nielsen E, Bichel P (1978) Acta Neuropathol. (Berl) 41:179

49) Mork SJ, Laerum OD (1980) J Neurosurg 53:198

50) Hirakawa K, Suzuki K, Ueda S, Nakagawa Y, Yoshino E, Ibayashi N, Hayashi K (1984) J Neurooncol 2:231

51) Wood JR, Green SB, Shapiro WR (1988) J Clin Oncol 6:338

52) Patronas NJ, Di Chiro G, Kufta C, Bairamian D, Kornblith PL, Simon R, Larson SM (1985) J Neurosurg 62:816

© 1989 Elsevier Science Publishers B.V. (Biomedical Division)
Cerebral gliomas. G. Broggi and M.A. Gerosa, editors

STAGING OF BRAIN GLIOMAS

MASSIMO SCERRATI, ROMEO ROSELLI and GIAN FRANCO ROSSI

Institute of Neurosurgery, Catholic University, Largo A. Gemelli 8,
00168 ROMA (Italy)

INTRODUCTION

In the year 1943-52 Pierre Denoix developed a system of staging classification of cancer called TNM (Tumor, Node and Metastasis), whose main principles were the applicability to all anatomical regions and the possibility of further modifications following surgical or pathological findings (post-surgical staging) (1,2). This system, firstly applied to breast and larynx cancers, is today used for 28 anatomical tumor areas among which cerebral tumor are not yet so far included. The TNM classification in fact hardly applies to brain tumors for many reasons: 1) the difficulty to define the true extension of the disease either radiologically or surgically; 2) the lack of lymphnodes involved by the tumor; and 3) the rarity of metastases, either intracranially or extracranially. Some attempts however have been made to stage intracranial tumors, in particular those of the pediatric age, such as PNETs of the posterior fossa (medulloblastomas) (3), craniopharyngiomas (4), germinomas (4,5), cerebellar astrocytomas (6), brain stem tumors (7) and optic pathway tumors (8). Nevertheless there is not at present time a suitable staging system for cerebral neuroepithelial tumors (9).

A contribution to reach this goal can be offered by stereotactic biopsy (10). The possibility in fact to obtain tissue samples from crucial sites of the body has been recently greatly favoured by the progress of the imaging radiology and has brought in oncology to the concept of "pathological staging". In neuro-oncology the capability of serial stereotactic biopsies to define the morphological as well as the spatial characteristic of brain tumors has been nowadays well proved (10,11,12,13,14,15,16,17,18).

We attempt here, analyzing the neuroradiological, pathological and surgical data of a series of 155 cerebral glial tumors, to propose a possible schema of staging classification.

MATERIALS AND METHOD

All the tumors of this series underwent stereotactic biopsy. Biopsies were planned on the basis of the previous neuroradiological investigations (CT scan and more recently MRI) in such a way to explore different part of the tumor as well as its periphery and the surrounding brain tissue (B.A.T.) (10,17,18,19,20). Eigthy-one turned out to be astrocytomas (5 grade 1, 50 grade II and 26 grade III), 20 were oligodendrogliomas (17 grade II and 3 grade III), 11 were oligo-astrocytomas (8 grade II and 3 grade III) and 43 were glioblastomas (grade IV). Surgical removal was carried out in 38 of the biopsied tumors. Radical removal was possible in 9 cases, tumorectomy in 11 and partial removal in 18. No discrepancy of histological diagnosis between bioptic and surgical specimes was observed in the cases operated upon.

PROPOSAL OF STAGING

To classify in stages brain gliomas we considered basically three aspects: the assessment of the extension of the neoplastic disease (T), the recognition of the biopathological characters i.e.histotype and degree of malignancy (G) and the evaluation of the extension of the surgical removal (S).

As far as the tumor extension is concerned, three cathegories are considered: size under 3 cm, between 3 and 5 cm, and more than 5 cm. This is due to the presumed different control yielded by therapy (surgery, conventional or focal radiotherapy).

Tumor grading is assessed according to the criteria of the W.H.O. classification (21,22), taking into account the modification suggested by L.B. Rorke et al. for the tumors of the pediatric age (23).

The extension of surgical removal is regarded as radical (1 cm beyond tumor borders, as for instance in lobectomies including the tumor) (Fig.1), tumorectomy (up to the extimated tumor borders) (Fig.2), and partial.

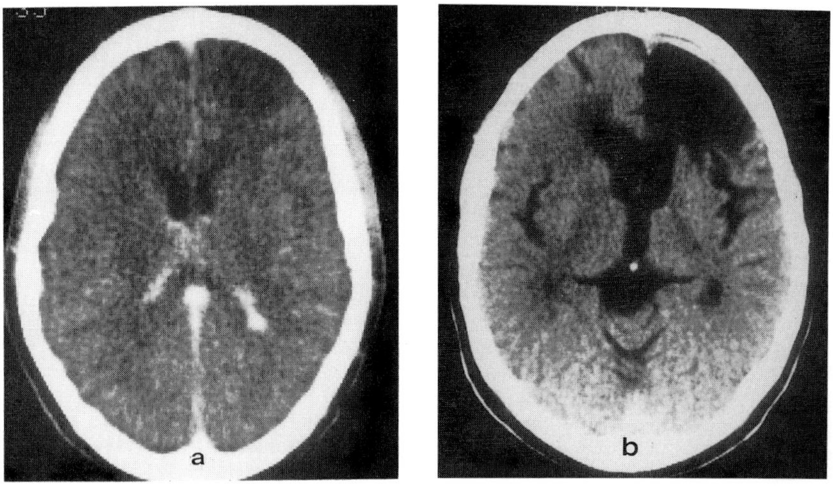

Fig. 1. Right fronto-polar astocytoma before (a) and after (b) radical removal.

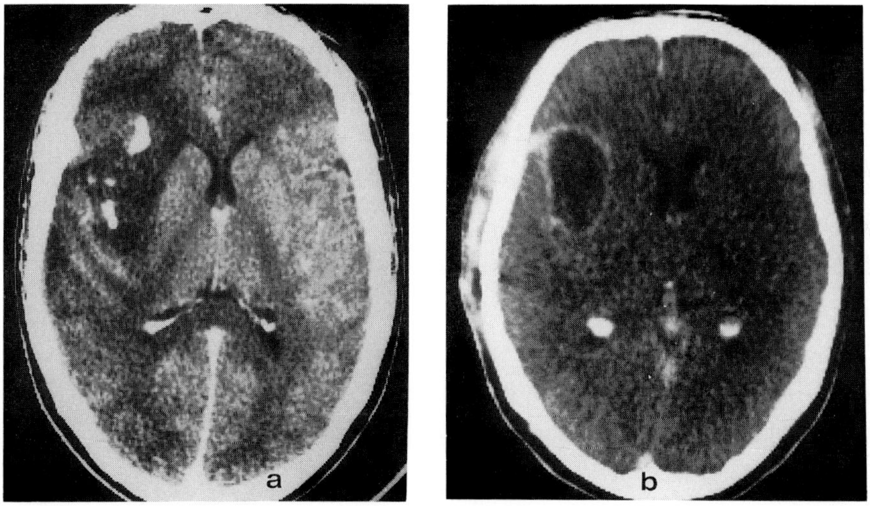

Fig.2. Left fronto-temporal oligodendroglioma before (a) and 1 week after (b) tumorectomy.

The proposed schema of staging for cerebral glial tumors is summarized in Table 1.

TABLE 1

STAGING OF CEREBRAL GLIAL TUMORS

SIZE	T_1	< 3 cm
	T_2	3-5 cm
	T_3	> 5 cm
GRADING	G_1	W.H.O. I
	G_2	W.H.O. II
	G_3	W.H.O. III
	G_4	W.H.O. IV
SURGICAL REMOVAL	S_1	radical
	S_2	tumorectomy
	S_3	partial

DISCUSSION

The classification in stages of evolution represents a way of objective description of the neoplastic disease through the different times of its natural history. It implies the definition of two essential components: the anatomical extent of the disease and the disease itself, i.e. the biopathological characteristic of the tumor. Its recognition contributes: a) to define the therapeutic strategy; b) to correctly express the prognosis; c) to control the effect of therapy; and d) to assess the validity of different therapeutic protocols on equivalent neoplastic diaseases.

Brain tumors, and neuroepithelial tumors in particular, present the peculiar characteristic to be in most of cases a focal and not a systemic disease. Recurrencesare in fact usually local and multicentric tumors are accounted no more than 3-6% of them (22,24,25,26). The main variables which influence the tumor evolution are in this context its anatomical extension, its biopathological characters and the extent of surgical removal.

The recent progress of the imaging neuroradiology (CT scan and MRI in particular) has greatly contributed to the direct identification of intracranial lesions and to the evaluation of their location and size. Nevertheless radiological means cannot properly define the true extension of the tumor (10,13,15,16). CT scan in fact often understimates the tumoral borders, while MRI can excessively enalarge them (16) (Fig.3). Stereotactic biopsy seems to us the only means able to define the extension of the neoplastic disease within the brain, provided that it be planned and carried out with multiple serial samples taken along tracks exploring the tumor and the surrounding brain as well (10,11,12,13) (Fig.4).

Neuroradiology, CT scan in particular, can sometimes suggest on the basis of the morphological characters the presumed malignancy of the tumor or of certain areas within it. No histological diagnosis can be done however on neuroradiological grounds only. Tumoral type and grading can be established only by pathological examination (10,13,15,16). The definition of the grading in neuroepithelial tumors is nowadays still a debated question. This is due mainly to the phenomenon of the progression which very frequently occurs in this tumors, the astrocytic one's in particular. Progression represents in fact the qualitative instability of the tumor during its natural history and its irreversible transition towards more heterogeneous stages. Its development within the tumor is usually focal and discontinuous, therefore its recognition is crucial for the understanding of its biological dignity. Stereotactic biopsy represents again, in our opinion, the most suitable means to detect this important aspect and to get a grading assessment as much correct as possible (10,16) (Fig.5).

Though the treatment of the neoplastic disease is to be regarded as multidisciplinary, surgery represents the most effective and immediate means to remove the tumoral mass or to reduce it (17,27,28). Moreover if we consider brain tumors mostly a focal disease, the role of the surgical removal in influencing the prognosis appears obviously of great relevance.

To conclude, the proposed schema wants to represent the first attempt to classify in stages of evolution the cerebral glial tumors. Large controlled trials aiming at verifying its correlation with the therapeutic outcomes and the prognosis are obviously necessary to prove its utility and its worth in neuro-oncology.

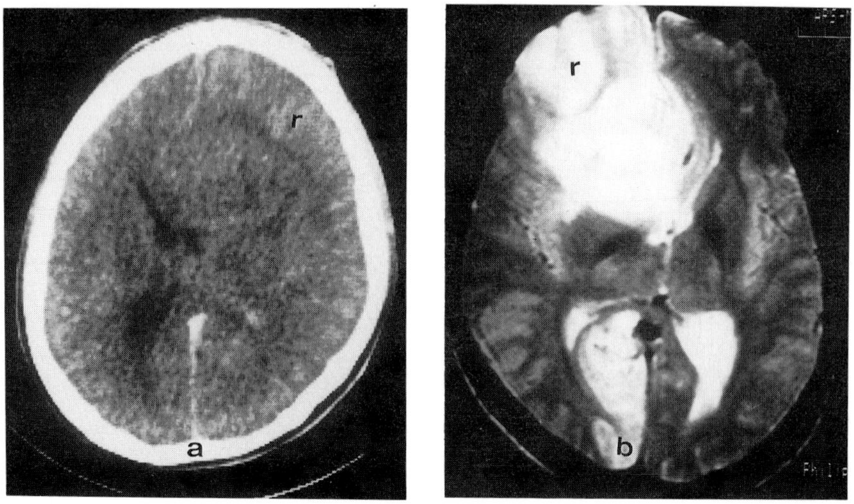

Fig. 3. CT scan (a) and MRI (b) of a frontal glioblastoma. Only indirect signs of mass effect are seen on CT scan, while a large expanding lesion involving both frontal lobes and the midline is shown by MRI.

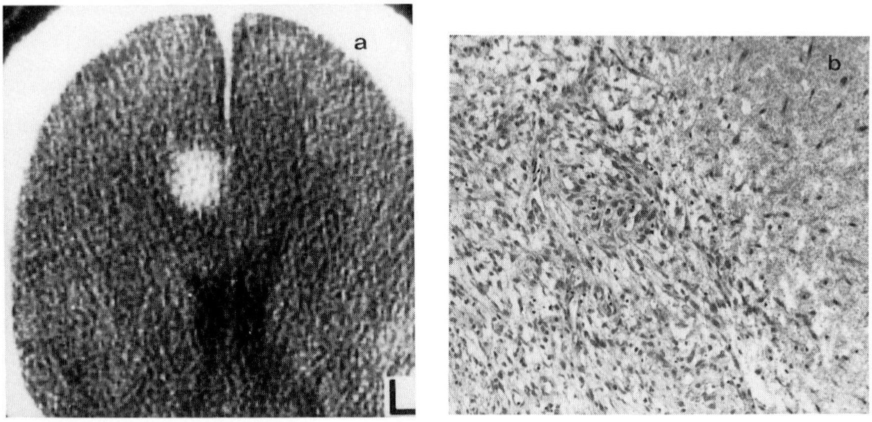

Fig.4. Pilocityc astrocytoma of the right gyrus cinguli. CT scan (a) and stereotactic specimen taken from the border of the tumor (b).

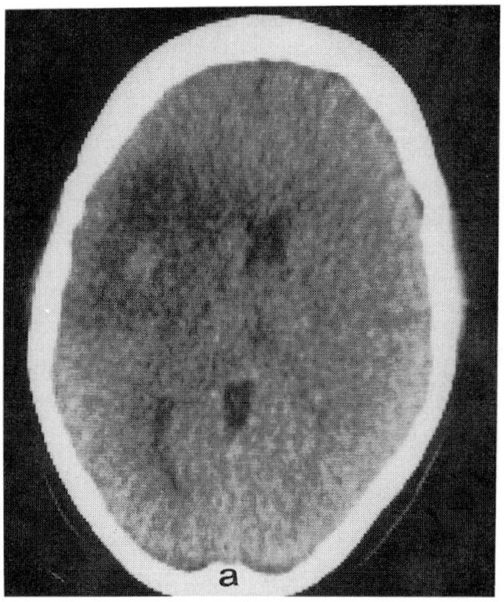

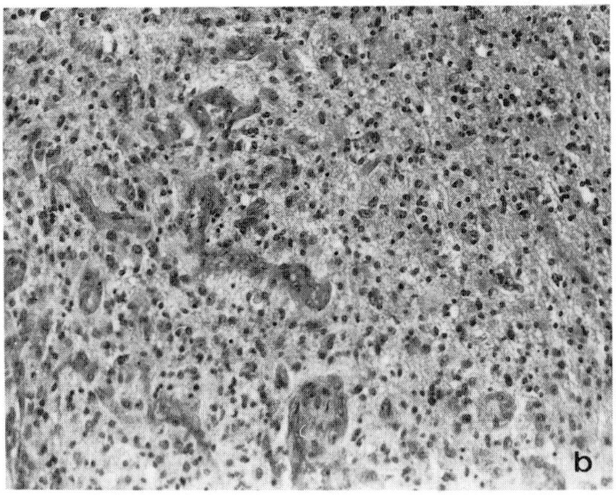

Fig. 5. Hyperdense area within a hypodense left fronto-temporal astrocytoma (CT sca, a) corresponding to a focal anaplasia within a more differentiated tumor (b) detected by stereotactic biopsy.

ACKNOWLEDGEMENTS

This research is partially supported by a grant of Italian Ministry of Education.

REFERENCES

1. Denoix PF (1944) Bull. Int. Nat. Hyg. 1:69
2. Denoix PF (1944-45) Bull. Int. Nat. Hyg. 52:82
3. Laurent JP, Chang CH, Cohen ME (1985) Cancer (Suppl)56 (7): 1807-1809
4. James HE, Edwards MsB (1985) Cancer (Suppl) 56(7): 1800-1803
5. Bruce DA, Allen JC (1985) Cancer (Suppl) 56(7): 1792-1794
6. Klein DM, McCullough DC (1985) Cancer (Suppl) 56 (7): 1810-1811
7. Epstein F (1985) Cancer (Suppl) 56 (7): 1804-1806
8. McCullough DC, Epstein F (1985) Cancer (Suppl) 56(7): 1789-1791
9. Edwards MSB, Klein DM (1985) Cancer (Suppl) 56(7): 1784-1785
10. Scerrati M, Rossi GF, Roselli R (1987) Acta Neurochir (Wien) (Suppl) 39: 28-33

11. Daumas-Duport C, Szikla G (1981) Neurochirurgie 27: 273-284

12. Daumas-Duport C, Monsaingeon V, N'Guyen GP, Missir O, Szikla G (1984) Acta Neurochir (Wien)(Suppl) 33: 185-194

13. Kelly PJ, Daumas-Duport C, Scheithauser BW, Kall BA, Kispert DB (1987) Mayo Clin Proc 62 (6): 450-459

14. Kleihues P, Volk B, Anagnostopoulos J, Kiessling M (1984) Acta Neurochir (Wien) (Suppl) 33: 171-181

15. Ostertag CB, Mennel HD, Kiessling M (1980) Surg Neurol 14: 275-283

16. Roselli R, Iacoangeli M, Scerrati M, Rossi GF (1988) Acta Neurochir (Wien) in press

17. Rossi GF, Scerrati M, Roselli R (1987) Appl Neurophysiol 50:159-167

18. Scerrati M, Rossi GF (1984) Acta Neurochir (Wien) (Suppl) 33:201-205

19. Scerrati M, Fiorentino A, Fiorentino M, Pola P (1984) J Neurosurg 61: 1146-1147

20. Scerrati M, Pola P, Fiorentino A, Fiorentino M (1987) Acta Neurochir (Wien) (Suppl) 39: 186-187

21. Zuelch KJ (1979) Types histologiques des tumeurs du systeme nerveux central. Classification,histologique internationale des tumeurs. OMS, Genege.

22. Zuelch CJ (1986) Brain tumors, Springer Verlag, Berlin.

23. Rorke LB, Gilles F, Davis RL, Becker LE (1985) Cancer 56: 1869-1985

24. Sauer R (1987) In: Jellinger K (ed) Therapy of malignant brain tumors. Springer, Wien New York, pp 195-276

25. Hochberg FH, Pruitt A (1980) Neurology 30:907-911

26. Russel DS, Rubinstein LJ (1971) Pathology of tumors of the nervous system, Arnold, London

27. Voth D. (1987) In: Jellinger K (ed) Therapy of malignant brain tumors. Springer, Wien New York, pp 91-126

28. Ransohoff J (1983) In: Walker MD (ed) Oncology of the nervous system, Martinus Nijhoff Publ., Boston, pp 101-115

TREATMENT

Cerebral gliomas. G. Broggi and M.A. Gerosa, editors

RECENT ADVANCES IN BRAIN TUMOR BIOLOGY

MARK ROSENBLUM, JAMES RUTKA, AND MICHAEL BERENS
Department of Neurological Surgery and Brain Tumor Research Center, School of Medicine, University of California, San Francisco, California 94143.

A greater understanding of brain tumor biology has led to better clinical care and a modest increase in survival of patients with brain tumors. However, a cure remains elusive. In this chapter, we summarize recent observations on the development and growth of brain tumors and speculate on possible new therapies.

ONCOGENESIS

When a normal cell's DNA is altered by spontaneous mutations or by exposure to radiation, chemicals, or viruses, excessive amounts of or alterations in the protein products of genes can lead to changes in cellular function.

Recent investigations have defined a set of normal cellular genes, usually those involved in the control of cell division (proto-oncogenes), that are found in excess quantities or in an altered state in tumors (putative oncogenes). If a homologous gene is found in a virus that can cause cells to express neoplastic behavior, a putative oncogene can be reliably defined as an oncogene (1-6). Several oncogenes govern the production of growth factors that can serve as a primary stimulus for mitosis or prepare a cell for a second factor that will promote cell division (7-9). Other oncogenes produce proteins that act as growth factor receptors at the level of the plasma membrane (10). Once an exogenous growth factor binds to a receptor, alterations occur in adjacent intracellular sites; this is usually associated with the activation of protein kinases and guanine triphosphate-binding proteins, and a biochemical cascade is initiated (1,2,5,6). Eventually, this cascade modifies intranuclear proteins and modulates DNA synthesis and cell mitosis (11).

Research on oncogenes and growth factors in human brain tumors is proceeding rapidly (12-14). The most common oncogene found in human brain tumors is erb-b, the epidermoid growth factor (EGF) receptor gene (15-18). The B chain of the platelet-derived growth factor (PDGF) is encoded by the sis oncogene; overexpression of this oncogene and secretion of PGDF-like proteins into the medium have been observed in glioma cell cultures (19). The myc gene has been found amplified and rearranged in a human glioblastoma cell culture (20), and a novel gli gene was recently identified in a malignant glioma (21). Finally, preliminary studies have shown excessive amounts of the N-ras gene in a majority of malignant gliomas biopsies and cell cultures (22); it has yet to be determined if N-ras is mutated or activated in those specimens. Research is now directed towards identifying cells that both possess growth factor receptors and produce the factor, thereby fulfilling the requisites for autocrine growth stimulation.

Additional oncogenes and protein products related to brain tumors are being sought by random screening of tumor specimens and by more direct approaches. Chromosome analysis of human gliomas has shown that clustering of chromosome alterations at specific break points is common (23). Since many oncogenes have been localized to specific chromosomes (24), a logical approach would be to search for genes on the chromosomes that are frequently abnormal. Structural chromosomal rearrangements could alter the gene product directly or cause the gene to be under altered regulation by changes occurring at a distance. Once the events important for the transduction of growth factor signals through the cytoplasm to the nucleus are better identified, specific means to interrupt this process can be found.

DETERMINANTS OF TUMOR GROWTH

Tumors are thought to originate from cells with altered growth regulation that proliferate exponentially unless retarded by environmental factors. Tumor growth can be considered as a balance between the proliferation and loss of tumor cells (Table 1). The proliferation of malignant cells is regulated by the clonal cell hierarchy, local oxygenation and nutrition, growth factors, and the ability of cells to migrate into the surrounding normal brain. Tumor cell destruction occurs as a result of inherent cell death, environmental factors, and immune mechanisms. The tumor cell population decreases as the dead cells are removed from the tumor site.

The balance between tumor cell proliferation and loss is a dynamic proces that is constantly changing in response to tumor size, regional vascular adaptations, and possibly host immune reactions. Without treatment, cellular proliferation usually predominates.

TABLE I

Determinants of brain tumor growth that result in increases or decreases in tumor cell number. Each determinant can be targeted by a treatment modality.

	Determinants of Tumor Growth	Potential Treatments
Increase in Tumor Cell Number	Clonal hierarchy	Differentiation agents
	Oxygen	Antiangiogenesis
	Nutrition	Metabolic inhibitors
	Growth factors	Growth factor inhibitors
	Cell migration	Migration inhibitors
Decrease in Tumor Cell Number	Inherent cell death	Radiation therapy, cytotoxins,
	Environmentally mediated	phototherapy, hyperthermia
	Immune mediated	Immune modulation
	Dead cell removal	Surgery

Factors inherent to the cell that are important in growth regulation include the clonal cell hierarchy, proliferative potential, and clonal expansion. The clonal cell hierarchy can be defined as the stratification of cells according to their growth potential from the original clonogenic stem cell. Some cells are capable of self-renewal and infinite cell division, while others have progressively less ability to proliferate and increasing degrees of differentiation. Cells with finite proliferative potential are considered "doomed" cells. Eventually, they cease to divide and from the standpoint of proliferation are considered "dead" cells, although they may continue metabolic activity or exist in a state of nonviable progressive cell lysis.

The proliferative potential of a cell is based upon its inherent kinetic properties. Cells not actively progressing through the cell cycle can either be permanently nonproliferating "dead" cells or "resting" clonogenic or doomed cells that are temporarily outside of the active cell cycle (G_0 phase). Resting cells can reenter the cycle under suitable local environmental conditions.

Clonal expansion occurs through the proliferation of a single clonogenic cell. The daughter cells may contain altered types or amounts of DNA that can be apparent as differences in chromosomes or cellular function (25). Over many cell divisions, surviving altered cells contribute to the heterogeneity of many malignant brain tumors. Additional factors, such as variable vascularization and the growth of a tumor into different areas of normal brain, also contribute to regional heterogeneity in individual tumors.

Tumor growth is regulated by several local environmental conditions. Without adequate oxygenation and nutrition, toxic products of cell proliferation and degradation accumulate and local tissue pH decreases. The amount of oxygen and glucose and regional pH can markedly influence cell metabolism and consequently cell division and migration. Growth factors secreted by infiltrating normal cells or by a subpopulation of tumor cells can stimulate tumor cell proliferation (paracrine growth mode) (25). A cell's proliferation can be decreased because of contact with adjacent cells or with surrounding extracellular matrix (ECM) macromolecules (26). The ECM surrounds normal brain cells and tumor cells and is interspersed between cerebrovascular elements and normal brain and tumor cells (27, 28). Interaction with the ECM can influence the growth rate and state of differentiation of a tumor cell. We have found that the ECM deposited by normal leptomeningeal cells grown in plastic tissue culture flasks similar to that surrounding normal blood vessels in situ and contains collagens types 1, 3, and 4, fibronectin, laminin, and hyaluronic acid. The proliferation of a tumor cell line (U 343 MGa) is inhibited when grown on that matrix (29). Other ECM macromolecules that bear study include glycoproteins, glycosaminoglycans and proteoglycans (27).

The influence of ECM proteins on cellular growth and differentiation probably involves cell membrane receptors and a balance between a cell's production of matrix-degrading enzymes and their inhibitors (29-31).

Recently, interest has been focused on angiogenesis as a prerequisite for the development of tumors. Lacking a proper blood supply, an incipient neoplasm cannot receive an adequate amount of oxygen and nutrients, which limits its growth (32). Angiogenesis factors can be separated into two broad categories: chemoattractants, which attract blood vessels into the tumor, and mitogens, which stimulate endothelial cell proliferation (32). The most important type of angiogenesis factor found in tissue cultures of human malignant brain tumors is the basic form of fibroblast growth factor (FGF).

Tumor cell migration is another important aspect of tumor growth. A tumor cell must not only lack contact inhibition, it also must have the potential for movement based on having the appropriate cytoskeleton and energy sources. It is likely that as yet unidentified environmental or cell-derived chemotactic factors direct tumor growth beyond the site of origin.

Two additional factors affecting tumor growth are the general nutrition of the host, which influences the proliferation of all cells of the body, and the host immune system. Immune modulation is mediated predominantly by lymphocytes and macrophages that infiltrate the tumor. Both types of cells can secrete cytokines, such as interleukin-1, interleukin-2, interferons, and tumor necrosis factor (TNF), which can influence tumor cell proliferation and differentiation (33). In addition, lymphocytes with natural-killer (NK) cell activity and lymphocytes that have been stimulated by IL-2 (lymphokine-activated killer cells, LAK) can recognize and kill tumor cells with which they come into contact.

Development of Future Treatments

Tumor growth should be considered as a balance between tumor cell proliferation and tumor cell loss; attempts to increase tumor cell loss or decrease tumor cell proliferation should inhibit tumor growth and decrease tumor size (Table 1). Methods that can decrease tumor cell proliferation include differentiation agents, antiangiogenic factors, metabolic inhibitors, growth factor inhibitors and migration inhibitors. Factors that can increase tumor cell loss include radiation therapy, cytotoxins, hyperthermia, phototherapy, attempts at immune modulation, and surgery.

Any single treatment is bound to have a limited effect on tumor growth because the balance of tumor cell proliferation and tumor cell loss is usually altered only moderately; under most circumstances, stabilization of the disease or a partial tumor response is all that can be expected. One alternative is to use combinations of agents that do not inhibit one another's actions and that are at least additive in their antitumor activity without compounding host tissue toxicity. The greatest hope for curing brain tumors lies in the development of better biological response modifiers and in the ability to detect tumor-specific and patient-specific sensitivities. If tumor vulnerabilities can be identified by studies of the growth factor-biochemical cascade important in growth regulation, new specific and potent antitumor agents might be developed.

ACKNOWLEDGMENT

Supported by grants CA 31882 and CA 13525 from the National Cancer Institute, and by a grant by the Preuss Foundation.

REFERENCES

1. Bishop JM (1983) Annu Rev Biochem 52: 301-354
2. Bishop JM (1985) Cancer 55: 2329-2333
3. Duesberg PH (1983) Nature 304: 219-226
4. Eva A, Robbins KC, Andersen PR, et al (1982) Nature 295: 116-119
5. Gordon H (1985) Mayo Clin Proc 60: 697-713
6. Lebowitz P (1983) J Clin Oncol 1: 657-663
7. Doolittle RF, Hunkapiller MW, Hood LE, et al (1983) Science 221: 275-276
8. Stiles CD (1985) Cancer Res 45: 5215-5218
9. Waterfield MD, Scrace GJ, Whittle N, et al (1983) Nature 304: 35-39
10. Downward J, Yarden Y, Mayes E, et al (1984) Nature 307: 521-527
11. Studzinski GP, Brelui ZS, Feldman SC, Watt RA (1986) Science 234: 467-470
12. Harsh GR, Rosenblum ML, Williams LT J Neurooncol (submitted)
13. Rosenblum ML, Harsh GR (1987) Modern Neurosurgery (in press)
14. Westermark B, Nister M, Heldin C-H (1985) Neurol Clin 3: 785-799
15. Filmus J, Pollak MN, Cairncross JG, Buick RN (1985) Biochem Biophys Res Com, 131: 207-215
16. Lieberman TA, Nusbaumtt R, Razon N, Kris R, Lar I, Soreq H, Whittle N, Waterfield MD, Ulrich A, Schlessinger J (1984) Nature, 313: 144-147
17. Lieberman TA, Razon N, Bartel AD, Yarden Y, Schlessinger J, Soreq H (1984) Cancer Res 44: 753-760
18. Westphal M, Harsh GR, Rosenblum ML, Hammonds RG Jr (1985) Biochem Biophys Res Comm 132: 284-289
19. Nister M, Heldin CH, Wasteson A, Westermark B (1984) Proc Natl Acad Sci USA 81: 926-930
20. Trent J, Meltzer P, Rosenblum ML, et al (1986) Proc Natl Acad Sci USA 83: 470-473
21. Kinzler KW, Bigner SH, Bigner DD, Trent JM, Law ML, O'Brien SJ, Wong AJ, Vogelstein B (1987) Science 236: 70-73

22. Gerosa MA, DellaValle G, Talarico D, Fognani C, Raimondi E, De Carli L, Tridenti G (1987) J Neurooncol 5: 172

23. Bigner SH, Mark J (1984) In: Rosenblum ML, Wilson CB (eds) Progress in Experimental Tumor Research, Vol. 27, Brain Tumor Biology. S. Karger, Basel, Switzerland, pp 67-81

24. Sparkes RS (1984) In: Bishop JM, Rowley JD, Graaves M (eds). Alan Liss, Inc, New York, pp 37-50

25. Sporn MB, Roberts AB Peptide growth factors and inflammation, tissue repair and cancer. J Clin Invest 78: 329-332

26. Nicolson GL (1987) Cancer Res 47: 1473-1487

27. Rutka JT, Apodaca G, Stern R, Rosenblum ML (1988) J Neurosurg 69: 155-170

28. Rutka JT, Myatt CA, Giblin JR, Davis RL, Rosenblum ML (1987) Can J Neurol Sci 14:25-30

29. Rutka JT, Giblin JR, Apodaca G, DeArmond SJ, Stern R, Rosenblum ML (1987) Cancer Res 47: 3515-3522

30. Apodaca G, Rutka JT, Banda MJ, Rosenblum ML, McKerrow, JH J Clin Invest (submitted).

31. Rutka JT, Apodaca G, McKerrow J, Giblin J, Rosenblum ML (1987) Proc AACR 28: 72

32. Shing Y, Folkman J, Sullivan R, Butterfield C, Murray J, Klagsbrun M (1984) Science 223: 1296-1299

33. Dinarello CA, Mier JW (1987) N Engl J Med 317: 940-945

Cerebral gliomas. G. Broggi and M.A. Gerosa, editors

CURRENT TREATMENT OF MALIGNANT SUPRATENTORIAL GLIOMAS

FRANCO PLUCHINO, SANDRO LODRINI, CESARE GIORGI and GIOVANNI BROGGI

Department of Neurosurgery-Istituto Neurologico-Milan-Italy

In 1926 Cushing said: "About gliomas,it is difficult to say anything intelligent" (1); nevertheless malignant gliomas are extensively operated in almost all neurosurgical departments even if postoperative survival is short and sometimes poor, so, facing a patient with glioma, we must answer some important questions:is surgery in these tumors always justifiable? and, if not, when have we to forbear?

We reviewed the last 103 patients operated in our Institute for a malignant supratentorial tumor with a minimum follow-up of one year (see Tab.1) and we matched our results with data from literature.

Tab.1.General characteristics of 103 patients with malignant gliomas

	N°	Mean age	Male/Female
Gbl	62	49(from 19 to 73)	38/24
Anapl. glioma	41	36(from 8 to 63)	25/16

A first important problem is the histological classification;extensive ROTG and ECOG (4) studies could not find a significant prognostic difference between III and IV W.H.O grades, on the contrary a clearcut difference in survival was noted between Glioblastomas (Gbl, with tumoral necrosis) and Anaplastic astrocytomas (without necrosis). Our series confirm this trend (Fig.1): at 12 and 24 months after the intervention the survival rate is 35 and 10% for Gbl and 90 and 80% for Anaplastic gliomas. A perfec but "pure histological" classification is, in terms of biological behaviour, less useful than a simpler but more practical one.

Another important prognostic element is the patients' age:about 80% of our patients aged over 50 years died within the 12 postoperative months while only 11% of the patients under 50 did . These data agree with BTSG (3) and ROTG (4) series : their mortality rates increase abruptly if the patients are over 50 (assuming 1 the mortality rate under 45,it rises to 1.75 over 55, to 3.5 over 65).

Patients' age is not however, an absolute prognostic factor: some long survivals have been observed even around the age of 60 in our series, and Gbl predominates over 50.

A third prognostic factor for Gbl that we noted in our patients is the type of removal: if it is partial,the mean survival is 32 weeks; if it is macroscopically total the mean survival is 52 weeks. Although denied in the past, the favourable effect of the total removal on survival seems to be sufficiently stated (12,13). Our series is not very large and we had 3 very long survivals in the group of "total" removal; besides, in the group of "partial" removal the early mortality

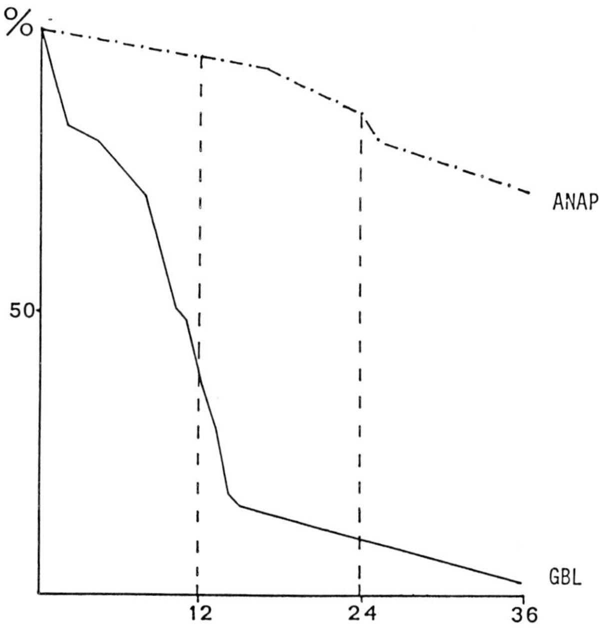

Fig.1 Difference in survival rate between glioblastoma (Gbl) and anaplastic glioma (Anapl)

is very high,reflecting the poorer conditions of these patients ; these facts may affect the difference in survival time . After the twelfth postoperative months, however, the two survival curves are parallel (Fig.1). Literature data show that

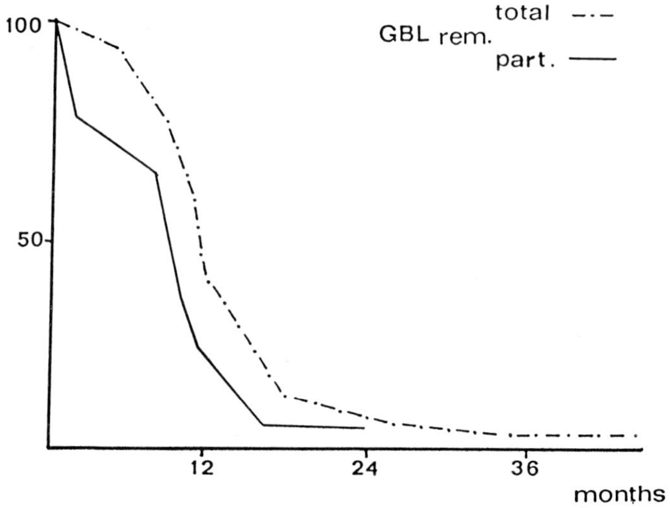

Fig.2 The influence of total or partial removal on survival time

postoperative radiotherapy is foundamental in prolonging the survival (4,8,17). Many efforts have been made to improve the radiation effects on the tumoral :mass hyperfractionation of the total dose (10,15), additional booster dose on the tumoral volume (4), use of "radiosensitizer" (9), but generally the result were not better than those obtained with the conventional schedule of 6000 rads given in daily dose of 200 rads/5 days per week. Late radionecrosis depends on the total dose and on the number of fractions ; it should be avoided if the dose/fractions ratio does not exceed the indications drawn from the literature by Sheline (14). All the patients in our series had radiotherapy except six, who died in the first six postoperative months. Despite the mass of published data on chemotherapy,its role in the treatment of malignant glial neoplasms is still controversial (7); the use of chloroetilnitrosureas is the most prominent and frequent and these compounds are also considered the most effective:carmustine and lomustine are the variant of the basic nitrosureas used in the clinical practice. Carmustine (BCNU) is the most extensively used and studied(4,6,17): it is deemed to improve the mean survival time from 40 weeks(surgery+radiotherapy) to 50 weeks (surgery+radiotherapy+BCNU). BCNU has howevwer an elevate mielo and epatotoxicity and can cause pulmunary fibrosis. Lomustine (CCNU) was found ineffective if used alone but it improves mean survival if associated to the radiotherapy(5,7).

We are using CCNU since 1978 ,because our double blind study(16) demonstrated better results compared to BCNU and CCNU is easily administrated by oral route, although it promote profuse vomit, so the real dose absorbed is questionable. Unfortunately, data regarding chemotherapy in our series presented here are not available. The efforts to improve the efficacy of the chemotherapy with other drugs (procarbazide, vincristine, teniposide, idroxiurea), alone or in association with nitrosureas, did not give positive results (7).

Recently a new surgical instrument has been developped at our Department. It is based on a graphic processor that allows the surgeon to use CT and MR images to plan the approaches to deep seated cerebral tumors.

Spatial stability between brain and skull during image acquisition and early operative phases makes it possible to reproduce at operation measurements and strategies worked out on CT and MR data. Spatial informations are all referred to a common reference system, a stereotactic frame worn both during image acquisition and at operation. Detailed description of the method is given elsewhere (C.Giorgi et al., this book). In synthesis the patient is under general anesthesia, the frame is positioned so that it does not interfere with a craniotomy, performed in the usual fashion. A CT or MR localizer identifies the image plane. Images are transferred to the surgical graphic processor by means of a 8" floppies, and, when the patient is taken to the operating room and prepared for surgery, the surgeon traces with a digitizer the contours of the lesion and other structures relevant for the surgical approach. The stuctures outlines, color coded, are then tridimensionally represented within a graphic outline of the stereotactic frame. The orientation of an ideal surgical probe can be selected on this image; angle and depth values ,corresponding to that same probe orientation on the stereotactic frame, are simultaneously indicated on the screen. The procedure is carried on with microsurgical technique. After a 2 cm cortical incision spatulas are gently advanced following the orientation of a thin catheter , which is stereotactically introduced up to the most superficial border of the lesion, aiming at its center. Stereotactic guide gives two

important advantages:
1) indicates the position of relevant cerebral structures that lie in proximity of the selected trajectory
2) allows to choose an approach according to the importance of brain areas surrounding the lesion, guiding the surgeon, when necessary, through longer paths.

This technique has undergone extensive clinical testing during the last 8 months, with constant intraoperative confirmation of the planned strategy.

DISCUSSION

The natural history of the malignant gliomas, namely of the Gbl, is very poor; if the patient is not treated, he will die in 2-3 months (12,13) while a large excision of the tumor allows a survival time of about 6 months(2). A mean survival time of 10-12 months is generally reached if radiotherapy is performed postoperatively. Operative mortality decreased from 25% to about 3% (12,13) in the last 20 years. This improvement is certainly due to the advances in anesthesia and medical treatment of intracranial hypertension,but also to the new operative techniques; so, microsurgery allows a minor cortical incision and better tumoral delimitation, ultrasonic aspiration (CUSA) and CO laser vaporization can further reduce peritumoral tissue damage.These facts led us to be generous in surgical indications for malignant gliomas;we must weigh patient's general conditions, his neurological status, tumoral location and the possible iatrogenic postoperative deficit. Our goal is to prolonge the survival time, leaving the patient in good neurological conditions. Only if surgical risk is too big ,we can consider palliative treatment(radiotherapy and chemotherapy). The new surgical facilities could,besides,change our therapeutical strategy about glial radiologically diagnosedand often histologically verified by means of stereotactic biopsy as low grade astrocytomas . Their surgica treatment is generally delayed until a consistent neurological deficit is present,i.e. sometime many years after a diagnosis of tumor has been made; generally, however, when these patients are operated on, the histological findings demonstrate a tumoral evolution toward malignancy,with a poor prognosis. Using all the new techniques, may be it would be better to operate low grade tumors before their malignant evolution.

Even if a great deal of studies are under way to improve interaction bet-ween the surgeon and the computer, it is clear to us that computer assisted microsurgery will soon change established criteria towards lesions considered inoperable because of their location. This is what is happening in our experience, where new perspectives would be opened by cases like the one shown in Fig.3

We all remember Cushing's sentence that a 3 years delay before the relapse af an operated Gbl is absolutely exceptional, but we observed in our practice some very long survivors ; four patients (one is not comprised in the present series) survived from 36 to 170 months after total removal of their glioblastoma (repeatedly checked by the neuropathologist because of the anomalous behaviour) and subsequent irradiation. There is then place for moderate hope for patients with malignant gliomas and we must use all available resources after a wise and careful observation of the patient's status.

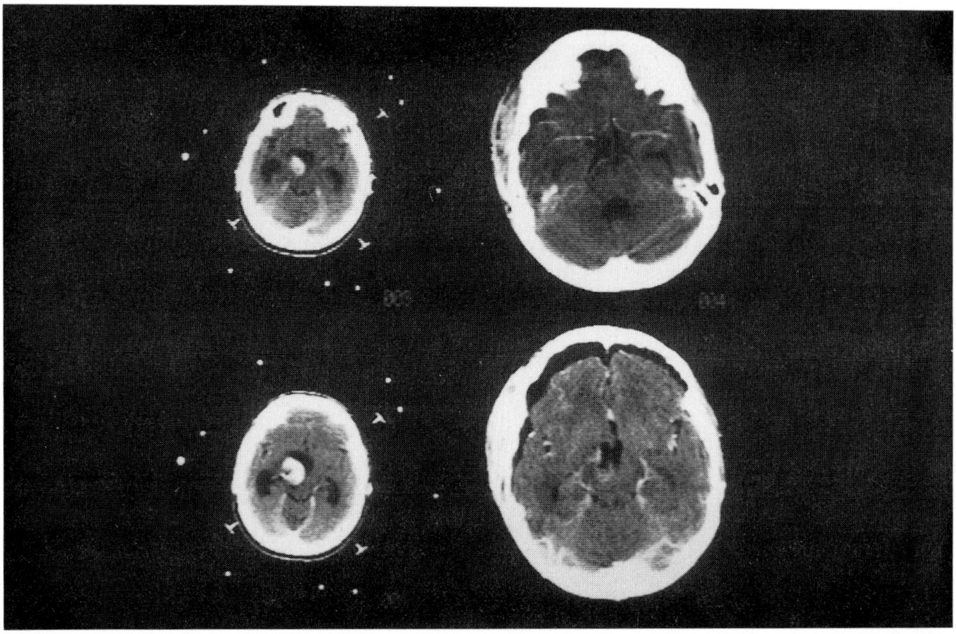

Fig.3 CT performed immediately before (left) and two days after stereotactically guided microsurgical removal of a thalamic fibrillary astrocytoma (right).These pictures are shown to demonstrate the efficacy of the method.

REFERENCES

1) Bayley P,Cushing H (1926)A classification of the tumors of the glioma group on a histogenetic basis with a correlated study of prognosis. 1 vol. Philadelphia,Lippincott

2) Blain JG,Guay JP,Derome G (1980) Un.Mèdic.Can 109:1-4

3) Byar DP,Green SB,Strike TA (1983) Prognostic factors for malignant gliomas in Walker MD ed. Oncology of the central nervous system Boston/The Hague pp 379-395

4) Chang CH,Horton J et al.(1983) A Joint Radiation therapy Oncology Group and Eastern Cooperative Oncology Group study Cancer 52 : 997-1007

5) Cianfriglia F,Pomili A,Riccio A et al.(1980) Cancer 45:1289-1299

6) Green SB, Byar DP, Walker MD et al. (1983) Cancer Treat Rep 67:121- 132

7) Kornblith PL,Walker M (1988) Chemotherapy for malignant gliomas 68:1-17

8) Leibel S,Sheline GE (1987) J Neurosurg 66:1-22

9) MRC Working Party (1983) Br J Radiol 56:673- 682

10) Payne DG,Simpson WJ et al. (1982) Cancer 50:2301-2306

11) Pecker J,Scarabin JM,Dekkiche M (1981) Neurochirurgie 27 287-293

12) Salcman M (1985) in: RH Wilkins,SS Reganchary(eds) Neurosurgery Mc Graw-Hill N.Y. Chapter 65 pp 579-593

13) Sheline GE, Wara WM, Smith V (1980) Int J Radiat Oncol Biol Phys 6:1215-1228

14) Shin H , Muller PJ, Geggie PHS (1983) Cancer 52:2040-2043

15) Solero CL, Monfardini S et al. (1979) Cancer Clin Trials 2:43-48

16) Walker MD, Strike TA, Sheline GE (1979) Int J Radiat Oncol Biol Phys 5:1725-1731

© 1989 Elsevier Science Publishers B.V. (Biomedical Division)
Cerebral gliomas. G. Broggi and M.A. Gerosa, editors

COMPUTER ASSISTED STEREOTACTIC PLANNING OF NEUROSURGICAL PROCEDURES

CESARE GIORGI, DAVIDE S.CASOLINO, ANGELO FRANZINI, GIOVANNI BROGGI AND FRANCO PLUCHINO

Neurosurgical Department, Istituto Neurologico "C.Besta" Via Celoria 11
20133 Milano.

INTRODUCTION

Recently introduced digital neuroimaging techniques, namely CT and MR, have raised new interest on stereotactic neurosurgery. High density and spatial resolution offered by these images can indeed provide the surgeon with all anatomical information necessary for approaching endocranial lesions.

Performing CT and MR under stereotactic conditions allows every image pixel to be referred to a surgical reference system. Since the brain can be safely assumed to maintain its position within the skullthroughout diagnostic phases until surgery, stereotactic approach to brain lesions can be planned, based on neuroanatomical information previously acquired.

Computer programs that allow for the selection of a target point on CT images and give the surgeon the angles and depth for a probe to reach it, are already available (1,2,3). This procedure is normally undertaken using CT or MR console, by examining the image sequence. Tridimensional arrangement of structures shown on CT or MR images and orientation of surgical trajectory are left to the surgeon's 3-D representation. The task is simpler when carried out or MR frontal, sagittal and semiaxial planes, but is still solved by looking at a series of parallel slices. Even if this solution is geometrically sound, and allows to reach any volume element of the brain, it doesn't provide in our opinion, necessary information for correct surgical planning.

Tracking structures encountered by the surgical probe in its approach to the target is left to the operator's ability to reconstruct tridimensionally the anatomical volume, since probe trajectory very seldom lies in one of the image planes.

Target is selected by examining successive sections, rather than the complete volume of the lesion: this constraint could probably only be accepted in case of a biopsy procedure; when surgery consists of placement of an isotope seed for interstitial radiation and particularly in case of guided removal of a lesion, detailed knowledge of spatial relationship among lesion and surrounding healthy brain cannot be ascertained by observation of successive biplanar anatomical images.

Furthermore, integration between CT, MR and angiographic data cannot be obtained since they are presented on separate radiological units.

Different authors have described methods to "frame" angiographic data within a 3-D stereotactic matrix, to integrate knowledge of vascular anatomy in planning surgical trajectories; again, in our opinion, they lack of the necessary capability of tridimensional representation (4,5,6).

In this paper we describe a surgical console, developed at Istituto Neurologico, to assist the surgeon in planning the approach to cerebral lesion, that has proved its value in extensive clinical practice.

MATERIAL AND METHODS

A Digital PDP 11/73 computer, driving two ECS Tesak EGP 414 and two high resolution graphic monitors has been chosen; it is integrated with an 8" floppy unit, that reads magnetic media common to all radiological unit in our Institute.

Programs are written in Fotran IV; different reconstruction algorithms have been implemented, to meet the geometries of available stereotactic frames. Decoding of floppies image formats has been carried out, to unable the surgical unit to read images from various radiological sources.

The surgical image processor is located in the operating room; a digitizer and a graphic monitor allow the surgeon to interact with images.

The patient undergoes suitable neuroradiological examination wearing a stereotactic frame, onto wich a localizer is mounted. This is an instrument of suitable material, that produces "artifacts" on the image, by means of which the image plane can be established, within the surgical frame of reference.

Normally, the exam is performed immediately before surgery, but it can be done several days before, if a device for the exact relocation of the frame is available.

Image transfer to the surgical console is completed in minutes, and images are presented on the graphic monitor during the preparation of the patient for surgery (FIG. 1).

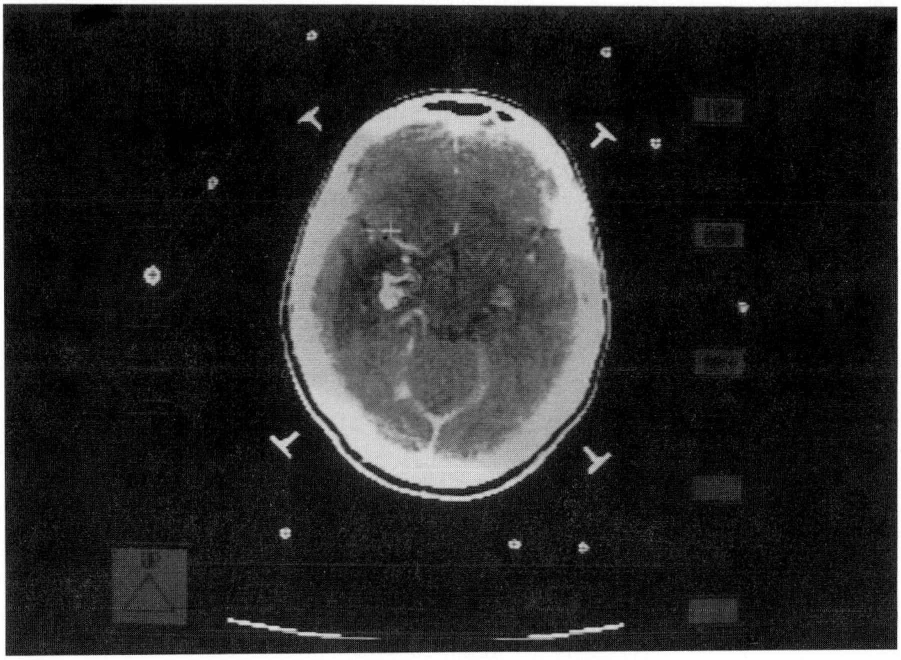

FIG. 1 : a CT image, from a series acquired preoperatively. Nine white dots are visible around the head: they are the artifacts produced by the localizer. Lesion outline and vessels are digitized with cursor (white cross).

The position of localizer artifacts on CT or MR is identified by the surgeon, by means of the digitizer, that controls a cursor on the screen, once the type of surgical frame in use has been specified.

After identification of the position of the image plane, the surgeon can trace contours of image elements relevant for surgery planning: using the digitizer, he outlines the lesion, the ventricles, major vessels and eloquent areas. This step is repeated for each CT or MR image; color codes identify borders of the same structure.

Once this acquisition process is completed, the computer draws a schematic outline of the frame in use, within which contours of structures are displayed. Around this picture, screen areas are allocated, dedicated to the control of image rotation and movement of surgical probe.

Angular and depth values, corresponding to the selected trajectory on the surgical frame, are listed on the screen.

Interaction with the graphic processor is facilitated by positioning the cursor, driven by the digitizer on the designated "command areas" displayed on the screen (FIG. 2).

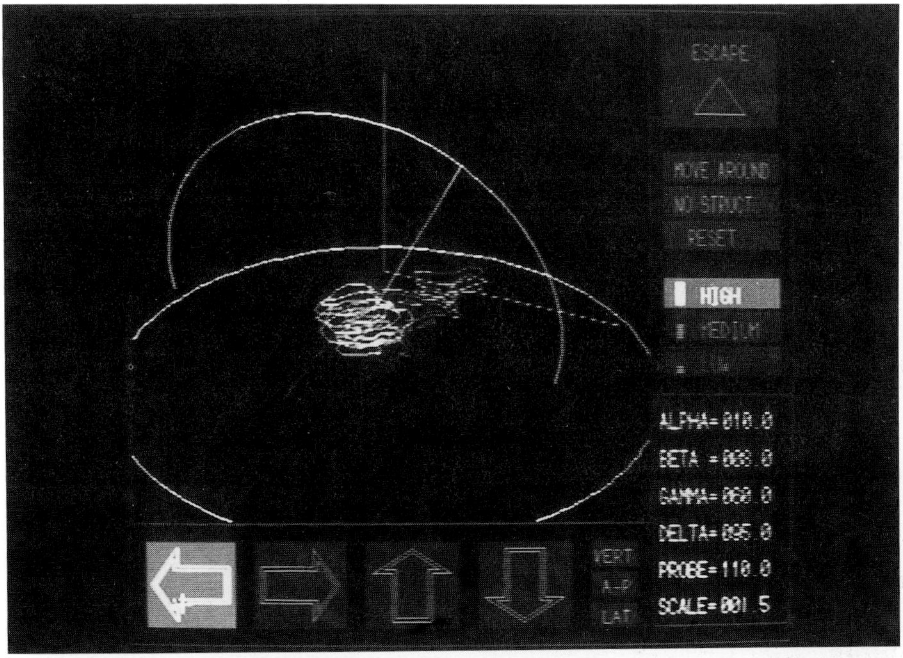

Fig. 2 : Parallel outlines of lesion, ventricles and major vessels, presented on a graphic monitor, within the geometry of a stereotactic frame. Image rotation and selection of probe trajectory are obtained by moving the cursor on peripheral "command areas" on the screen. Angles and depht corresponding to the chosen trajectory are indicated in lower right corner of the screen.

A different procedure is followed to reconstruct vascular anatomy: it is based upon the acquisition of multiple stereoscopic angiographic pairs, acquired under stereotactic conditions. Images are obtained with a producion digital angiographic unit, slightly modified to hold the stereotactic frame in the isocenter of the C arm carrying the X Ray tube and to allow a given sequence of stereoscopic pair. A specially designed localizer carries three small steel spheres, mounted at different heights along the perimeter of the frame. These spheres project shadows on the sequence of stereoscopic angiographies whose position is digitized by the surgeon, once that the images have been transferred to the surgical graphic console.

The position of the spheres identifies univocally the acquisition angle, since distance between XRay focal spot, stereotactic frame and image plane are fixed and known, as is the frame position with respect to the plane of the C arm carrying the XRay tube. The computer can then calculate the projection of a stereotactic trajectory and superimpose it over every corresponding angiographic image. The surgeon selects the probe trajectory by examining CT or MR images. This trajectory is then superimposed to the angiography stereoscopic pairs, thus allowing the tridimensional perception of the probe orientation within vascular branches (FIG. 3).

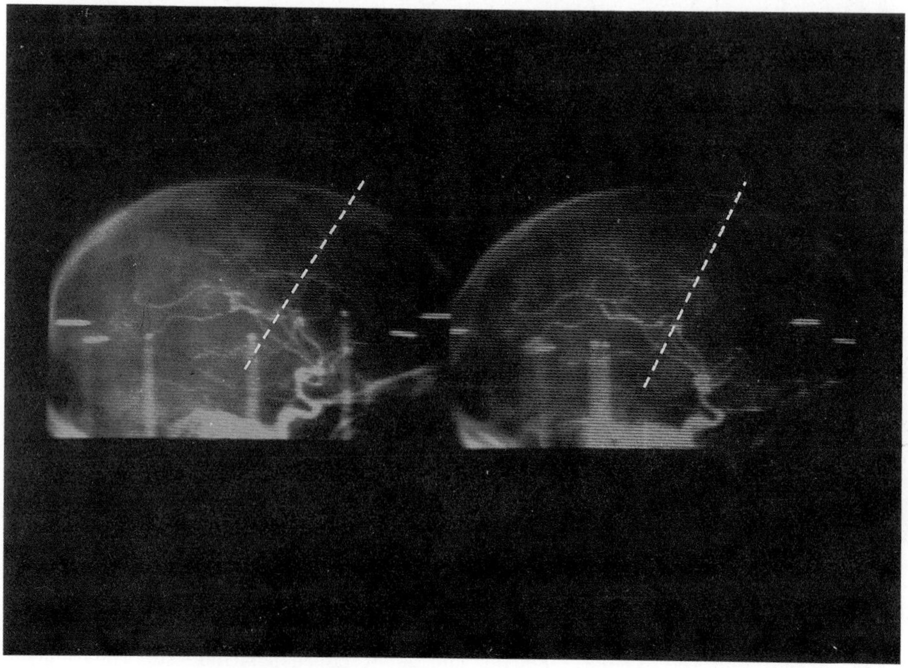

Fig. 3 . Two stereoscopic angiographies, on which the computer has superimposed the probe position, projected in corresponding angular views.

RESULTS

Experimental accuracy tests, carried on at our laboratory show a precision in target localization within +or- 1mm in the plane of CT images. Accuracy falls within +or-2mm,in case of data reading along the axis orthogonal to the image plane, whenever "thin slicing" algorythms are used.

Even if these limits could be theoretically improved, we think that actual precision is adequate for clinical applications:we have performed over 30 cerebral biopsies using this method, obtaining specimen from the areas selected on CT in all cases.

Accuracy of the method using MR data is still questionable and is material of present research. Normally MR information is superimposed to CT images to assist the selection of targets for biopsy.

The great potential of this method has prompted us to use it as a guide for cerebral lesion removal. In its actual stage of development, the system allows to stereotactically place a thin probe in the parenchima, reproducing the trajectory planned with the 3-D reconstruction. Microsurgical removal can be carried out, following this method, through a 2cm cortical incision, with a custom retractor.

In cases operated upon with this technique, the possibility to reach deep seated lesions with limited trauma to the surrounding normal brain has been proved.

Simultaneous display of structures of surgical relevance, like major vessels, the internal capsule, diencephalic nuclei, and cortical eloquent areas, allows for planning of trajectories that spare such structures,rewarding even when imply a longer surgical path.

The inevitable shift in brain volume following craniotomy, has never caused significant errors in target identification.

Lesion removal has always been carried out with microsurgical technique. $CO2$ laser and ultrasonic aspirator have been extensively employed, together with custom designed stereotactic retractors and micro instruments, to allow the surgeon to work in narrow and deep surgical areas.

DISCUSSION

As largely reported in literature (7,8,9), there is a general trend towards the development of methods that allow brain lesions removal with minimal surgical trauma, using CT stereotactic guidance. In our experience the great potential offered by this technology promotes the development of research projects in the field of neurosurgical instrumentation and intraoperative monitoring devices.

At present the development of autostatic retractors and "on line" guiding devices that eliminate the need of cumbersome stereotactic arcs, is under way. We also feel the need for intraoperative monitoring support, particularly during advanced operative phases, when relationship between structures is likely to be altered by tissue removal and CSF drainage.

Adequate development of support to this technique will contribute to the diffusion of stereotactic concepts in the domain of microsurgery, thus improving its efficacy.

REFERENCES

1. Apuzzo MLJ, Sabshin JK (1983) Neurosurgery 12:277-285

2. Mundinger F, Birg W, Klar M (1978) Appl. Neurophysiol. 41:169-182

3. Lunsford LD, Martinez AJ, Latchaw RE (1986) J. Neurosurg. 64:872-878

4. Bergstrom M, Greitz T, Ribbe T (1986) Neuroradiology 28:100-104

5. Peters TM, Clark JA, Olivier A, Marchand EP, Mawko G, Dieumegarde M, Muresan LV, Ethier R (1986) Radiology 161:821-826

6. Kelly PJ, Alker GJ, Kall BA, Goerss S (1984) Neurosurgery 14:172-177

7. Sedan R, Peragut TC, Farnarier PH, Derome PH, Fabrizi A (1988) Neurochirurgie 34:97-101

8. Shelden CH, Mc Cann G, Jacques S, Lutes HR, Frazier RE, Katz R, Kuki R (1980) J. Neurosurgery 52:21-27

9. Kelly PJ, Alker GJ (1981) Surg. Neurol. 15:331-334

Cerebral gliomas. G. Broggi and M.A. Gerosa, editors

THE STEREOTACTIC BIOPSY OF BRAIN LESIONS : A CRITICAL REVIEW

Claudio MUNARI (1, 2), Osvaldo Oscar BETTI (3)

(1) : INSERM U 97, 2 ter rue d'Alésia, 75014 PARIS, FRANCE

(2) : Centre Hospitalier Sainte Anne, Service de Neurochirurgie, 1, rue Cabanis, 75014 PARIS, FRANCE

(3) : Istituto Medico Antartida, Avenida Rivadavia 4990, Buenos Aires, Argentine.

INTRODUCTION

There is an apparent general agreement concerning the need of knowing the precise diagnosis of an intracranial lesion before to decide the most appropriate treatment.

Several clinical arguments can suggest the existence of an intracranial lesion, but its location can be only suspected and cannot be firmly defined without the contribution of other informations allowed, in the recent past, by the EEG and the neuroradiological studies (DAVID, 1967)[1].

The recent utilization of the computed tomography (CT scan and Magnetic Resonance Imaging (MRI) strongly improved the possibility of early discover and localize the existence of an intracranial "lesion", even if its dimensions are very small. We voluntary use here the term of lesion, immediately raising the problem of the definition of a lesion.

In fact, the CT scan can only show, and correctly visualize, the existence of an altered density in the brain. On the other side, it is well known that the altered signal revealed by the Magnetic Resonance does not have, up to now, an always precise and well established significance.

This is to say that, in many cases, also considering all the available data fournished by the clinical history, neurological symptomatology, EEG, arteriography, CT scan, MRI), the diagnosis of the type of the lesion may remain unclear or, at least, not very precise.

However, the diagnosis of the type of the lesion, indispensable priorly to the therapeutic decisions, is only a part of a more complete diagnosis also considering the volume, the evolutivity, the spatial organization, the functional relationships and the sensitivity to the therapeutic managements of a given lesion.

Many of these questions may rapidly and efficaciously be answered when the lesions are located in superficial areas of the brain : the direct surgical approach and the complete removal, when it is possible, was and remains the best solution

When the lesion is small, deeply located or in highly functional areas, with not well defined limits, the direct surgical approach may be difficult and, over all, its therapeutics advantages are less evident that the encountered risks.

SOME HISTORICAL INFORMATION

To reach deep brain regions without to open the skull became possible using stereotactic methodologies directly derived from the Horsley and Clarke apparatus employed in animal experiments[2]. It was rapidly clear that the osseous reference points, very useful in animals, did not allow the high precision indispensable in humans.

This is the reason for the original creation of stereotactic Atlases permitting to localise different brain areas on the basis of intracranial reference structures[3,4,5,6,7,8,9,10].

The initial success of this new and original approach to the brain was tremendous : during the "first International Symposium on Stereo-Encephalotomy (Stereotaxic Surgery)" held in Philadelphia in 1961, E.A. SPIEGEL[11] affirmed : "Fourteen years have passed since I suggested to Dr. WYCIS to apply the stereotaxic technique to the human brain...

... about three dozen guiding apparatuses have been developed...

... over 5 000 operations have been reported in the world Literature since 1948 when he performed the first stereotaxic pallidotomy"[12].

The maximum interest of the stereotaxic neurosurgeons was initially oriented to three types of neurological diseases : the dyskinesias, the pain[13] and the epilepsy.

In patients with abnormal movements, the aim was to cure the clinical symptomatology interrupting some specific pathways or destroying some nuclei.

In patients with epilepsy the first stereotactic investigations did concern the basal ganglia with the aim of improve the knowledge of the "centro-encephalic" epileptic syndromes (the "primary generalized epilepsies" of to day).

In spite of the very large number of stereotaxic operation, it is only in 1955 that the use of stereotactic biopsies is mentioned with a clear diagnostic aim, as well in hypophysis tumours (TALAIRACH et TOURNOUX, 1955)[14], as in brain gliomas (TALAIRACH et al., 1955)[15]. The conclusion of this last paper is : "Nous insistons également sur la nécessité de prélever du tissu par aspiration à travers la sonde et de pratiquer ainsi un examen histologique extemporané". Anyhow, we must wait 1960 for finding the first papers concerning the methodology of the Stereotaxic Biopsies (S.B.)

TABLE I

STEREOTACTIC BIOPSY

First papers reporting stereotactic cerebral biopsies

	PTS	YEAR
HOUSEPIAN AND POOL [16]	28	1960
HEATH ET AL. [17]	?	1961
JINNAI ET AL. [18]	?	1961
HOUSEPIAN AND POOL [19]	?	1962
HALLAC AND BACKAY [20]	3	1963
KALYANARAM AND GILLINGHAM [21]	62	1964
OTTOLENGHI [22]	24	1966
(CO-AUTHORS NASHOLD)		

The first reported cases are of patients with abnormal movements, and only in 1963 HALLAC and BAKAY[20] publish 3 stereotactic biopsies in patients having intracranial intracerebral tumours. Their original instrument was tested, before its use in neurosurgical patients, "in fresh cadaver brains and in experimental animals", and it was "used with good results and without any complication for liver biopsy in over 50 patients".

Answering to a direct question very recently given to him by the Authors of this paper, concerning the lack of scientific papers concerning the problem of stereotactic biopsies,at that time, Professor TALAIRACH said that the scientific problem was to be able to identify and to reach different cerebral structures and not to aspirate some fragment of brain tissu, that is relatively easy...(personal communication, november 1988)[23].

METHODOLOGICAL PROBLEMS

It seems today evident that a stereotactic methodology allows a safer and easier facility of obtaining cerebral samples. This apparently obvious concept was matter of controversy since free hand biopsy specimen, obtained with probes gave, in some cases, very high rate (93 %) of correct cytological classification[24]

The good results obtained with a sterotactic methodology were not immediately able to overcome the difficulties due to the neurosurgeons reluctant to perform apparently "blind" biopsies in the brain. And this in spite of the fact that the S.B. seem less dangerous that the electrocoagulation : KALYANARAM and GILLINGHAM, in 1964, report 8 bleedings on 388 electrocoagulations, and 0/155 stereotactic biopsies [21].

BOSCH, in 1980 [25], trying to define the indications for S.B. in brain tumours, consider that the need of a precise histological diagnosis did occur with the improvement of the informations allowed by the neuroradiological investigations: "previously, only the relatively large mass lesions could be demonstrated... and craniotomy was only carried out when surgeons considered the lesion to be at least partially removable. Histological diagnosis formerly was of minor importance in inaccessible mass lesions as other forms of treatment (i.e., radiation and chemotherapy) were not yet established".

He states that the best indications are :

- midline, all size tumours,
- impossible complete or bulk resection,
- multiple intracranial mass lesions,
- single intracerebral mass lesion in patients with a known primary tumour, but without signs of metastasis elsewhere.

The S.B. were only performed in 60 cases of supratentorial tumours (26.7 % of a two years population), avoiding biopsies in infratentorial region. He considers that the use of the spiral needle is too hazardous in the brainstem, but admits that aspirative biopsies may be indicated in some cases.

In the same period, EDNER [26] published the results of a serie of 345 S.B. performed by the Karolinska Hospital Neurosurgical Staff during the 1968-1977 period, using the Leksell apparatus and the Backlund methodology [27].

We choose these 2 papers as good examples exemplifying the problematic approach to the S.B. in the early eighties- which can thus be summarized :

- the CT scan allows the discovery of smaller brain lesion ;

- the CT scan, stereotactically used, helps to reach the target ;

- the cerebral angiography (though not peroperative) shows the vascularity of lesions[25], but

- the rare bleeding complications (5/345) "would be avoided if a stereotactic angiogram had been included in the biopsy procedure" [26].

- The mortality rate,though appreciable seems acceptable (3 % of BOSCH cases, less than 1 % in the EDNER series) considering the inaccessible lesions selected for this approach.

It appears clear that the use of computer "created a renaissance in stereotactic surgery" (KALL, 1987)[28]. This author strongly stress that the stereotactic surgery was born in 1947, only one year after the first electronic calculator, but that "these fields would not intersect until years later" [28].

In fact, two different ideas of the stereotactic procedures are at least in partial opposition.

1 - On one side, an intracranial lesion may be considered as a target to be reached. It is necessary to visualize the target or its apparent aspect : alterations in density, altered signal at the MRI, or, also, mass effect appreciated

with "classical" neuroradiological investigations. Once the evidence of an abnormality is obtained, the computer can provide the useful informations indicating the trigonometrical coordinates for reach the abnormality-target in the stereotactic space.

This attitude may be called "ballistic" since the final result is to identify a track going from an entry point on the skull (in general, preconaric) to a given intracranial point. The geometrical precision of the CT scan coordinates is obviously of great help.

2 - On the other side, an intracranial lesion may be considered as inserted in a complex anatomical and functional world, as it is the brain. This approach does imply that the anatomy of the individual brain must be known before considering the lesion. This anatomical and, often,"functional" abnormality must be localised, of course, in the intracranial space, but principally considering its anatomo-functional context. Accepting this methodological approach, it becomes evident that the choice of a trajectory for performing S.B. should consider not only the arrival point, but also all the structures through which the biopsy needle will go.

These two views are at the origin of stereotactic frames which are completely different. In spite of the apparences, it is no matter about the geometrical forms of the frames, the differences being strictly linked to the underlying anatomical views. Thus, the SPIEGEL and WYCIS apparatus[12] is a spheric one, while the Talairach's frame[3,4] is rectangular.

But the important difference,also considering the biopsy problem, is that the Talairach's frame allows an orthogonal approach to the intracranial structures, respecting the anatomical plans.

Besides, the Scerrati's instrument[29] coupled with the Talairach's frame, allows to perform angular trajectories (more easily than the "Stereometre" of Sedan[30] while the Brown Roberts Wells may be now used with grids permitting an orthogonal approach[31]

However, the geometrical modifications of the frames and of accessory pieces, do not necessarily correspond to a metho dological evolution of the basic ideas.

The more and more frequent indications for a S.B. made desiderable the use of an universal, unic frame, permitting the best approach for every case.

NEURORADIOLOGICAL INVESTIGATIONS AND STEREOTACTIC BIOPSIES

Many authors use only the diagnostic data furnished by the CT scan, while others consider the angiographic data useful[32] or even mandatory[33,34].

In our group, at the Ste Anne Hospital, we consider that the stereotactic approach of an intracranial lesion must take into account not only the localization of the lesion, but also the individual anatomy of a given patient. Thus the stereotactic methodology of Talairach[3,4,7,8,9,10] is not simply resumed in the utilization of a stereotactic frame, but needs the identification of the cerebral structures surrounding the lesion, and of those the biopsy tracks will go through. The stereotaxic and stereoscopic angiography[35] allows the identification of the sulci and of the cerebral convolutions, thus permitting both, to choose the best adapted trajectory for the biopsy needle, and to avoid risks of cortical and subcortical bleeding.

The absence of severe side effects due to the angiography (slight transitory hypotension in 2 % of 134 children)[36] seems encouraging, also considering the quality of the obtained informations.

The ventriculography is also matter of controversy : several authors do it in all patients[37] while some others consider it useful only for lesions near of the ventricular system[38] or that the useful of the ventriculography stopped with the CT scan[39].

For us[40], the ventriculography aim is not limited to show the possible alterations of the ventricular system : the utilization of the proportional Talairach's stereotactic methodology[7,9,10] does permit to identify all the cerebral structures in a given brain. Moreover, the ventriculography data improve the manual reconstruction of the previous CT scan data in the stereotactic coordinates[41]

Despite the very high relevance of the data of the stereotactic neuroradiologic investigations ("positive" in 97 % of a series of 134 young patients)[36] it is evident, in our mind, that such investigations do not authorize to establish a precise histological diagnosis.

In the same study[36], we show that there is a very strong variability of the CT scan features, in contrast with many Authors considering that the CT scan data are enough for defining the therapeutic management, without considering the relatively high percent of misdiagnosis (14.5 % of 314 cases in the KENDALL series [42].

However, the CT scan data may only hardly be considered as providing reliable prognostic informations : PIEPMEIER considers that low grade hemispheric astrocytic tumors have a significantly longer survival when they are not contrast enhanced (and when the age is lower). Conversely, STROINK et al.,[44] consider that pediatric brain stem gliomas have a better prognosis when they are isodense and contrast-enhanced.

These being the literature controversies, we perfectly agree with HOOD et al.[45,46] considering that a correct histological diagnosis is mandatory for both, deciding the therapeutic management, and, later, evaluating the obtained results.

Thus, up to day, we perform the peroperative stereotactic and stereoscopic angiography and ventriculography with the double aim of better anatomically defining the lesion and of avoiding bleedings, as stressed by EDNER [26].

Our manual reconstruction of the CT scan data in the stereotactic space [41] implies slight errors. We will be, very soon, able to do stereotactic CT scan, but, up to now, we think that the <u>peroperative angiogra phy</u> will not be taked out of the operative protocol.

Conversely, it is possible that the MRI, obtained in stereotactic conditions, will made unnecessary the peroperative ventriculography, at the condition that the geometric distortion becames correctable.

Magnetic Resonance Imaging is considered to have greater sensitivity than CT scan for discovering intracranial lesions [47] However, the use of an effective contrast agent, with strong paramagnetic properties like the Gadolium-labeled DIETHYLENE-TRIAMINE PENTAACETIC ACID (D.T.P.A.), does not seem to be able "to depict tumor boundaries" or "to allow differentiation between tumor and radiation necrosis" (EARNEST et al., 1988)[48] . Moreover, these authors documented, by stereotactic biopsies,"isolated tumor cells beyond the margins of enhancing tumor tissue, and these cells extended beyond the lesion margins that were noted on T2-weighted MR images".

They conclude that because of the inability of CT scan and MRI to determine tumor margins, biopsy will remain an essential means of lesion assessment.

OPERATIVE STEREOTACTIC SURGERY

In our practice[49,50], almost all stereotactic procedures are done under general anaesthesia, using the Talairach's frame. The burr holes are done with a percutaneous drill of 2.45 mm diameter.

The large majority of the biopsy tracks are oriented normally to of the three plans of the space, but the Talairach's inclinable grids are also used, as well as the

Scerrati's arc for the twofold oblique tracks.

The serial biopsies are done with the Sedan Vallicioni instrument, partially modified : the external diameter varies between 2. 4 and 1.6 mm ; the lenght of the aspiration window varyes between 4 and 9 mm.

The number of tracks, and of obtained specimen is decided in agreement with the diagnostic and the therapeutic perspectives. Thus, in candidates to an interstitial curietherapy [51] or to the radiosurgery [52], the biopsies are performed not only in the center of the lesion, but also at its periphery and even, when it is possible, in the neighbooring areas.

On the other hand, in patients with lesions like cryptic vascular malformations [53], or localized in highly functional areas as in the brainstem [54], the number of samples is voluntary reduced to the minimum.

Various kinds of biopsy instruments are used by the different surgical groups[27,30,55],but also by the same group according to the characteristics of the lesion to be biopsied.

In our experience, the absolute number of tracks and of the obtained specimen progressively diminished from 1979 to 1986, as well as the number of tracks/intervention, the number of tracks/patient, and of specimen/patient.

However, since the mean number of specimen/track is absolutely identical in two compared periods (1979-1982 and 1983-1986)[36], what decreased was the number of tracks/patient

Many reasons can explain this change :

- The improvement of CT scan reconstruction ;

- A more selective choice of patients really needing a precise "spatial definition" ;
- The evolution of the diagnostic and therapeutic protocols.

This aspect of the stereotactic approach is difficult to be discussed, since in only a few studies, we found these data : BOUVIER et al. (1983)[56] did a number of tracks/intervention higher than our (2,4), while their number of specimen/intervention was lower (3,35). OSTERTAG obtains "usually three to six specimens"[57].

The aspiration of the brain tissue is preceded, in highly functional areas, by the measurement of the electrical impedence[58,37] , and by the electrical stimulation (1/sec ; 0.1-0.3 msec ; 0.4-2mA).

Every specimen is immediately divided in two parts for both, peroperative examination and after paraffine imbedding[59,60] The histological data are classified according to the OMS classification,[61] partially modified by DAUMAS-DUPORT.

Several authors, not giving the number of obtained speci men in different patients, stress the importance of obtaining biopsies of different (low and high attenuating) CT scan areas, thus improving the quality of the obtained diagnostic informations.[62]

More recently, it was demonstrated that MRI guided biopsies can allow a more precise definition of the volume and of the spatial configuration in glial tumours.[63]

Finally, it is evident that the way indicated by the

M.N.I. group of Montreal[64] is the best, since they can integrate the data furnished by :

- the digital angiography ;
- the Magnetic Resonance ;
- the CT scan ;
- the Positrons Emission Tomography (P.E.T. scan).

Some specific operative problems must and can be solved in patients with very deep seated lesions, as orbital regions[65] or brain stem or posterior fossa lesions[45, 54, 66, 67, 68, 69, 70].

CONTRIBUTION OF THE STEREOTACTIC BIOPSY TO THE THERAPEUTIC MANAGEMENT

In many cases, the stereotactic biopsy simultaneously allows the diagnosis and the treatment.

The best exemples are obviously represented by the non tumoral lesions :

- colloid cysts may be diagnosed and aspirated at the same time [71, 72]
- intracerebral hematomas can be diagnosed and partially evacuated[73]
- brain abscesses may be identified[74] and also cured[75] with a stereotactic methodology.

The particular problem of the obstructive hydrocephalus can be solved by a stereotactic biopsy of the floor of the IIIth ventricle[76,77].

The differential diagnosis between tumoral and non tumoral lesions is not always very easy, since the CT scan findings are only rarely pathognomonic of a precise lesion. This diagnostic difficulty assumes a particular importance in children.

RIEGEL et al. (1979) [78] in a recent review of papers published between 1932 and 1979, concerning the management and the prognosis of the brain tumors in children, stressed that the importance of a diagnosis prior to External RadioTherapy (ERT), or other therapeutics, is strongly underevaluated, while the survival curves are very well studied.

The need of a precise diagnosis appears much more evident considering that the ERT as well as the cytostatic therapy may be responsible of severe undue effects [79,80,81].

Several studies show that encephalic localized lesions may be non tumoral in a not negligible percent of cases : 18 % of 904 intracranial lesions stereotactically explored by MUNDINGER (1985) [82] ; 17 % of 261 stereotactic biopsies done by KELLY (1986) [83] ; 17 % of 180 cases of MUNARI et al., (1987 [50] 14.7 % of 401 cases explored by MORINGLANE and OSTERTAG (1987) [84] 13 % of 1236 cases of OSTERTAG [57], 12.6 % of 134 consecutive children explored at the Ste Anne Hospital [85].

Admitting that an ERT should not be done before a precise diagnosis, the problem raises of which is the best method for obtaining such a diagnosis.

The open surgery, realized with a purely diagnostic aim, does not seem the best solution, since the morbidity and the mortality are relatively high [86,87], and the diagnostic results themselves are often deceptive[44].

The results of the S.B. can offer possibilities to stereotactically treat several kinds of lesion. The interstitial irradiation of brain tumors was attempted since the early fifties[88]. In 1955, TALAIRACH and his co-workers published their first results obtained with the stereotactic implantation of radioactif Au 198 in 5 patients showing a rapid and astonishing improvement in 3 of them[15]

Because of technical difficulties linked to the utilization of the Radium, the Iridium 192 proposed by HENSCHKE in 1956[89], was firstly used in intracerebral lesions by the group of Presbyterian Hospital in New York[90,91].

The results obtained in very large series of patients[49,51,82,92,93] suggest that this kind of therapy has its role to play in the management of deep brain tumors with a maximum diameter of less than 30mm.

Craniopharygiomas represent a very interesting exemple of intracranial pathology, very difficult to treat by classical neurosurgical tools, and then benefiting of the diagnostic and therapeutic possibilities affered by the stereotactic methodology[94,95].

In fact, this "baffling problem" (CUSHING, 1932)[96] can not be always solved by the radical surgical extirpation.

The cystic component can be identified and its volume can be precisely defined with a stereotactic approach[97].

Further results did confirm that the beta endocavitary irradiation can produce an important reduction or even the disappearence of the cyst, a well with Yttrium 90 as with Rhenium 186[89,99,100]. Moreover, this kind of treatment is also useful in some cases of low grade glioma cysts or pseudocysts[99].

Concerning the solid craniopharyngiomas, after the stereotactic confirmation of the diagnosis, a stereotactic radiosurgical treatment can be done[101, 102].

Similarly, the management of lesions of the pineal region, of difficult and dangerous surgical approach should be firstly diagnosed by stereotactic biopsies[55,103,104].

In this way, the surgery can be only reserved for the non radiosensitive lesions.

An original computer-assisted stereotactic system was recently proposed for using the Laser in stereotactic conditions in deep seated intracranial lesions[69,83] : with this methodology, the S.B. become a therapeutic tool and a direct surgical guide during the open surgery.

It becomes more and more frequent to discover small brain lesions in patients with severe, drug resistant, partial epilepsies : the S.B. allow to precisely define the volume and the nature of the leion, thus modulating the strategy of the stereo-EEG explorations[105,106,107].

More over, this approach permits to improve the knowledge on the spatial and functional relationships between the "lesional" area (which does not necessarily strictly correspond to the "tumor") and the "epileptogenic" area(s), giving informations on the electrical activity <u>in</u> the tumors, at its periphery and in surrounding cerebral areas [108,109]. The final open surgery of the epilepsy is programmed considering all the collected data [110].

SIDE EFFECTS

The mortality rate linked to stereotactic procedures is very low : 3 % in the studies of BOSCH (1980)[25] and of OSTERTAG et al., (1980)[55], less than 1 % in the EDNER Series and no mortality in most of the published series[56,70,111,112].

A preexistant deficit may be definitively impaired in a very small percent of cases, as 1.4 % of 134 children operated on at the Ste Anne Hospital [36].

Similar results are reported in recently published studies in children [113,114].

These results are not very different from those well known in adults, varying from 0 % of the Rennes group [115] to 308 % of PECKER et al., 1979 [116], 5.9 % of LUNDSFORD and MARTINEZ (1984)[117].

One may consider that a non diagnostic biopsy is an undesirable effect : the rate of these not useful investigations is very low, varying from 0 of BROGGI et al., 1983[111]; 1.4 % of SCERRATI and ROSSI (1984)[118], 1.5 % of MORINGLANE et al., (1987)[113], 2.2 % of MUNARI et al., (1987)[50], 3.8 % of GODANO et al.,[114] (1987). OSTERTAG (1988)[57] states, after the revision of the literature that "depending on the physician's experience the overall rate of obtaining and accurate diagnosis by stereotactic technique exceeds 90 %".

However, we agree with SCERRATI and ROSSI (1984), stating that "the reliability of stereotactic biopsy seems related to the modality of its planning and execution and to the experience of the neuropathologist".[118]

CONCLUDING REMARKS

There is a general agreement on considering the stereotactic biopsies as an useful and safe neurosurgical tool for improving the management of intracranial lesions.

In our view, the diagnosis of a brain lesion should answer questions concerning :

- nature
- real volume
- spatial organization
- funtional relationships
- evolutivity

of a given lesion in an individual patient.

The used frame does not seem to be, at the present time, a real problem, considering the technological evolution allowing all the possible approach ways with many stereotactic frames.

The neuroradiological investigations, prior to S.B., are differently evaluated by different Authors, but the peroperative Angiography seems, for us, indispensable, at least in perisylvian pineal, IIIth ventricle lesions.

The CT scan and the MRI informations increase our knowledge on the location and the extension of lesions, without be able, up to now, to take the place of S.B. for the spatial definition and, above all, for histological identification.

The recent development of the stereotactic neurosurgery is opening new perspectives for the treatment of many kinds of intracranial pathology.

To obtain specimen of intracranial lesions increases our possibility of better treat many types of tumoral pathology, avoiding the risks of treat benign non tumoral lesions as the tumoral ones.

We believe that the stereotactic methodology allows new trends in CNS research, going from the development of experimental tumoral models [119] to the tentati-ve treatment of Parkinsonism by autologous transplantation of cathecholamine tissue [120,121].

More over, the stereotactic methodology can modify some "traditional" treatment of wall known pathology (e.g., the stereotaxic reconstruction of the aqueduct of Sylvius [122];

Stereotactic biopsies do not need any more favourable comment since almost all available data show that it is a safe and useful procedure.

The next step should probably be to develop new stereotactic solutions of old neurosurgical problems thus realizing that BACKLUND calls "a dream for clinical neuroscientists since the first days of Neurosurgery..." : "Surgical repair of the CNS" [121].

REFERENCES

1. DAVID M (1967) Neurochirurgie 13:181-205

2. HORSLEY V, CLARKE RH (1908) Brain 31:45-124

3. TALAIRACH J, HECAEN H, DAVID M, MONNIER M, AJURIAGUERRA J de (1949) Rev Neurol 81:4-23

4. TALAIRACH J, AJURIAGUERRA J de, DAVID M (1950) La Presse Médicale 38:697-701.

5. SCHATELBRAND G, BAILEY P (1959) Introduction to stereotaxis with an atlas of the human brain. Thieme Stuttgart.

6. SPIEGEL EA, WYCIS HT (1952) Sterero-encephalotomy (thalamotomy and related procedures). Part I. Methods and Stereotaxic Atlas of the Human Brain. Monographs in Biology and Medicine I. Grune & Stratton, New-York.

7. TALAIRACH J, DAVID M, TOURNOUX P, CORREDOR H, KVASINA T. (1957) Atlas d'anatomie stéréotaxique des noyaux gris cen traux. Masson et Cie. Paris.

8. TALAIRACH J, DAVID M, TOURNOUX P (1958) L'exploration chi rurgicale stéréotaxique du lobe temporal dans l'épilepsie temporale. Masson et Cie. Paris.

9. TALAIRACH J, DAVID M, TOURNOUX P et al. (1967) Atlas d'anatomie du télencéphale. Masson et Cie. Paris.

10. TALAIRACH J, TOURNOUX P (1988) Co-planar stereotaxic. Atlas of the human brain. 3-dimensional proportional system : an approach to cerebral imaging. Thieme, Stuttgart.

11. SPIEGEL EA (1962) Confin Neurol, 22:170.

12. SPIEGEL EA, WYCIS HT, MARKS M, LEE AJ (1947) Science 106:349-350.

13. HECAEN H, TALAIRACH J, DELL MB (1949) Rev Neurol 81:917-931.

14. TALAIRACH J, TOURNOUX P (1955) Neurochirurgie, 1:127-131.

15. TALAIRACH J, ABOULKER J, RUGGIERO G, DAVID M (1955) La semaine des Hôpitaux de Paris 31:1-6.

16. HOUSEPIAN EM, POOL JL (1960) J.Nerv & Ment Dis 130:520-525

17. HEATH RG, JOHN S, FOSS O (1961) Arch Neurol (Chicago) 4:291-300.

18. JINNAI D, NISHIMOTO A, MATSUMOTO K, HANDA S (1961) Excepta Medica 36:E94-E95.

19. HOUSEPIAN EM, POOL JL (1962) Confin Neurol 22:171-177.

20. HALLAC I, BAKAY L (1963) J Neurosurg XX:529-530.

21. KALYANARAM S, GILLINGHAM FJ (1964), J Neurosurg XXI:854-858.

22. OTTOLENGHI

23. TALAIRACH J (1988) Personal communication

24. MARSHALL LF, ADAMS H, DOYLE D, GRAHAM DL (1973) J Neurosurg 39:82-88.

25. BOSCH DA (1980) Acta Neurochir 54:167-179.

26. EDNER G (1981) Acta Neurochir 57:213-234.

27. BACKLUND EO (1971) Acta Chir Scand 137:825-827.

28. KALL BA (1987) Appl Neurophysiol 50:9-22.

29. SCERRATI M, FIORENTINO A, FIORENTINO M, POLA P (1984) J Neurosurg 61:1146-1147.

30. SEDAN R, DUPARET R (1968) Neurochirurgie 14:577.

31. APUZZO MLJ, FREDERICKS CA (1988) In: Lundsford LD (ed) Modern Stereotactic Neurosurgery, Martinus & Nijhoff, Boston, pp 63-77.

32. LOBATO RD, RIVAS JJ, CABELLO A, ROGER R (1985) Appl Neurophysiol 45:426-430.

33. GAHBAUER H, STURM V, SCHLEGEL W, PASTYR O, SCHARFENBERG H SCHABBERT S (1988) Amer J Neuroradiol 4:715-718.

34. KELLY J JR, ALKER GJ, KALL BA, GOERSS S (1984) Neurosurg 14:172-177.

35. SZIKLA G, BOUVIER G, HORI T, PETROV V (1977) Angiography of the human brain cortex. Atlas of vascular patterns and stereotactic cortical localization. Springer, Berlin, Heidelberg, New-York.

36. MUNARI C, ROSLER JR, MUSOLINO A, FRANZINI A, DAUMAS-DUPORT C, MISSIR O, CHODKIEWICZ JP (1988) J Ped Neurosc

37. BROGGI G, FRANZINI A (1981) J Neurol Neurosurg Psych 44:397-401.

38. NECKSTARD P, SORTLAND O, HOVIND K (1982) Neuroradiol 33:85-88.

39. WENDE S, KISHIKAWA T, HUWEL N, KAZNER E, GRUMME T, LANKSCH W (1982) Neuroradiol 23:85-88.

40. MUNARI C, MUSOLINO A, DAUMAS-DUPORT C, MISSIR O, BRUNET P GIALLONARDO AT, CHODKIEWICZ JP, BANCAUD J (1986) Appl Neurophys 48:440-443.

41. MUSOLINO A, MUNARI C, BETTI OO, LANDRE E, BROGLIN D, DEMIERRE B, MISSIR O, DAUAMS-DUPORT C, CHODKIEWICZ JP (1987) Rev EEG Neurophysiol Clin 17:11-24.

42. KENDALL BE, JAKUBOWSKI J, PULLICINO P, SYMON L, (1979) J Neurol Neurosurg Psychiat 42:485-492.

43. PIEPMEYER JM (1987) J Neurosurg 67:177-181.

44. STROINK AR, HOFFMAN HJ, HENDRICK EB, HUMPHREY RP (1986) J Neurosurg 65:745-750.

45. HOOD TW, STEPHEN MD, GEBARSKI MD, KEEVER PE, VENES JL (1986) 65:172-176.

46. HOOD TW, VENED JL, McKEEVER PE (1986) J Pediat Neurosurg 2:79-88.

47. BRADLEY WGJr, WALUCH V, YADKLEY RA, WYCOFF RR (1984) Radiology 152:695-702.

48. EARNEST FN, KELLY PJr, SCHEITHAUER BW, KALL BA, CASCINO TL, EHMAN RL, FORBES GS, AXLEY PL (1988) Radiolology 166:823-827.

49. SZIKLA G, BLOND S, DAUMAS-DUPORT C, MISSIR O, MIYAHARA S, MUNARI C, MUSOLINO A, SCHLIENGER M (1983) It J Neurol Sci suppl 2:83-96.

50. MUNARI C, MUSOLINO A, DEMIERRE B, BETTI OO, FRANZINI A, ROSLER JR, BROGLIN D, DAUMAS-DUPORT C, MISSIR O (1987) Rev EEG Neurophysiol clin 17:3-10.

51. SZIKLA G, PERAGUT JC (1975) Neurochirurgie 21:187-228.

52. BETTI OO, ROSLER JR, GALMARINI (1988) in the same book

53. DAUMAS-DUPORT C, MANN M, MUNARI C, BLOND S, MUSOLINO A, MONSAINGEON V, CHODKIEWICZ JP (1986) Appl Neurophysiol 48:440-443.

54. MUNARI C, MUSOLINO A, ROSLER JR, BLOND S, DEMIERRE B, BETTI OO, DAUMAS-DUPORT C, MISSIR O, CHODKIEWICZ JP, (1987) Appl Neurophysiol 50:200-202.

55. OSTERTAG CB, MENNEL HD, KIESSLING M (1980) Surg Neurol 10:513-520.

56. BOUVIER G, COUILLARD P, LEGER SL, LESAGE J, ROTENT F, BEIQUE RA (1983) Appl Neurophysiol 46:695-702.

57. OSTERTAG CG (1988) In: Lundsford L.D.(ed), Modern Stereotactic Neurosurgery, Martinus & Nijhoff, Boston. pp 63-77.

58. BENABID AL, ROUGEMONT J de, PERSAT JC, BARGE M, CHIROSSEL JP (1978) Neurochirurgie 24:3-14.

59. DAUMAS-DUPORT C, MONSAINGEON V, SZENTHE L, SZIKLA G (1982) Appl Neurophysiol 45:431-437.

60. DAUMAS-DUPORT C, MONSAINGEON V, BLOND S, MUNARI C, MUSOLINO A, CHODKIEWICZ JP, MISSIR O (1987) J Neuro-Oncol 4:754-757.

61. ZULCH KJ (1979) OMS (ed).

62. BOETHIUS J, COLLINS VP, EDNER G, LEWANDER R, ZAJICEK J, (1978) Acta Neurochir. 40:223-232.

63. KELLY P.Jr, DAUMAS-DUPORT C, KISPERT DB, KALL BA, SCHEITHAVER BW, ILLIG JJ (1987) J Neurosurg 66:865-874.

64. OLIVIER A, PETERS TM, CLARK JA, MARCHAND E, MAWKO G, BERTRAND G, VANIER M, ETHIER R, TYLER J, LOBTINIERE de A, (1987) Rev EEG Neurophysiol Clin 17:25-43.

65. OSTERTAG CB, UNSOLD R, WEIGEL K (1983) Neuro-Ophtalm 3:277-280.

66. APUZZO MLJ, SABSHIN JK (1983) Neurosurgery 12:277-285.

67. GALANDA M, NADVORNIK P, SRAMKA M, BASANDOVA M (1984) Acta Neurochir. 33:213-217.

68. COFFEY RJ, LUNSFORD LD (1985) Neurosurgery 17 : 12-18.

69. KELLY P.Jr, KALL BA, GOERSS BS (1986) Surg Neurol 25:530-534.

70. FRANZINI A, ALLEGRANZA A, MELCARNE A, GIORGI C, FERRARESI S, BROGGI G (1988) Acta Neurochir 42:170-176.

71. BOSCH DA, RAHN T, BACKLUND EO (1978) Surg Neurol 9:15-18.

72. MUSOLINO A, MUNARI C, FOSSE S, BLOND S, BETTI OO, DAUMAS DUPORT C, CHODKIEWICZ JP (1987) Appl Neurophysiol 50:210-217.

73. BACKLUND EO, VON HOLST (1978) Surg Neurol 9:99-101.

74. WALSH PR, LARSON SJ, RYTEL MW, MAIMAN DJ (1980) Appl Neurophys 43:205-209.

75. BROGGI G, FRANZINI A, PELUCHETTI D, SERVELLO D (1985) Acta Neurochir. 76:94-98.

76. POBLETE M, ZAMBONI R (1975) Confin Neurol 37:150-155.

77. MUSOLINO A, SORIA V, MUNARI C, DEVAUX B, MERIENNE L, CONSTANS JP, CHODKIEWICZ JP (1988) Neurochirurgie in press

78. RIEGEL DH, SCARFF TB, WOODFORD JE (1979) Child's Brain 5:329-340.

79. DICKINSON WP, BERRY DH, DICKINSON L et al (1978) J Ped 92:754-757.

80. ROWLAND J, GLIDEWELL O, SIBLEY R, HOLLAND JC, BRECHER M, TULL B et al (1982) Asco Abstracts

81. DUFFNER PK, COHEN ME, THOMAS PRM, LANSKY SB (1985) Cancer 56:1941-1946.

82. MUNDINGER F (1985) Acta Neurochir (Wien) 35:70-74

83. KELLY P.Jr (1986) Neurol 36:535-541.

84. MORINGLAND JR, GRAF N, OSTERTAG CG (1987) Rev EEG Neurophysiol clin 17:45-53.

85. MUNARI C, ROSLER JR, MUSOLINO A, BETTI OO, DAUMAS-DUPORT C, MISSIR O, CHODKIEWICZ JP (1988) Acta Neurochirur, in press.

86. FRANKEL SA, GERMAN WJ (1958) J Neurosurg 15:489-503.

87. HITCHOCK E, SATO F (1964) 21:497-505.

88. SACHS E (1954) J Neurosurg 11:119-121

89. HENSCHE VK (1956) Proc Int Conf Peaceful used of atomic energy 10:48.

90. CHASE NE, ATKINS HL, CORRELL JW (1961) Radiology 77:842-843

91. CORRELL JW, CHASE NE, ATKINS HL (1961) J Neurosurg 18:800-803.

92. MUNDINGER F (1970) In: Wang Y, Paoletti P (eds) Radionuclide applications in neurology and neurosurgery pp 199-265.

93. MUNDINGER F (1988) In: Schmided HH & Sweet WH (eds) Chapter 43 in operative neurosurgical techniques, Grune & Stratton, Inc. pp 491-514.

94. LEKSELL L (1951) Acta Chir Scand 102:316-319.

95. LEKSELL L, LIDEN K (1952) In:London Her Majesty's Stationery Office, Radioisotope Techniques.Vol1 Medical and physiological applications p 76.

96. CUSHING H (1932) In: Thomas (ed) Intracranial tumors, Springfield

97. LEKSELL L, BACKLUND EO, JOHANSSON L (1967) Acta Chir Scand 133:345-350

98. MUSOLINO A, MUNARI C, BLOND S, BETTI OO, LAJAT Y, SCHAUB C, ASKIENAZY S, CHODKIEWICZ JP (1985) Neurochirurgie 31:169-178.

99. MUNARI C, MUSOLINO A, BETTI OO, CLODIC R, ASKIENAZY S, CHODKIEWICZ JP (1988) In: Pluchino F & Broggi G (eds), Advances Technology in Neurosurgery, Springer Verlag, New-York, Berlin, Heidelberg, pp 120-131.

100. MUNARI C, LANDRE E, MUSOLINO A, TURAK B, CHODKIEWICZ JP (1988) It J Neurosurg Sci, In press

101. BACKLUND EO (1969) In: Hamberger CH & Wersäll J (eds) Nobel Symposium 10 : Disorders of the skull base region. Almqvist & Wiksell, Stockholm p237.

102. BETTI OO, DERECHINSKY VE (1984) Acta Neurochir 9:385-390

103. BACKLUND EO, RAHN T, SARBY B (1974) Acta Radiol 13:368-376.

104. BETTI OO (1983) La Semana Medica ano XC, n°5270, Tomo 163, n°13 : 1-9

105. BANCAUD J, TALAIRACH J, GEIER S, SCARABIN JM (1973) In: Edifor (Ed), EEG et SEEG dans les tumeurs cérébrales et l'épilepsie, pp 1-351

106. MUNARI C, TALAIRACH J, MUSOLINO A, SZIKLA G, BANCAUD J, CHODKIEWICZ JP (1983) It J Neurol Sci, 2:69-82.

107. MUNARI C, BANCAUD J (1985) In: Morselli PL & Porter RJ (eds), The Epilepsies, Butterworths, London, pp 267-306.

108. MUNARI C, MUSOLINO A, BLOND S, BRUNET P, GIALLONARDO AT, CHODKIEWICZ JP, BANCAUD J (1986) In: Schmidt D., Morselli PL (eds), Workshop on Intractable Epilepsy. Experimental and Clinical Aspects, Raven Press, New-York, pp 129-146.

109. MUNARI C (1988) In: Broggi G (ed), The rational basis of the surgical treatment of epilepsies, John Libbey Eurotext, London, Paris, pp 121-138.

110. TALAIRACH J, BANCAUD J, SZIKLA G et al (1974) Neurochir. 1:240pp

111. BROGGI G, FRANZINI A, MIGLIAVACCA F, ALLLEGRANZA A (1983) Child's brain 10:92-98.

112. ROSSI GF, SCERRATI M, ROSELLI R, (1987) Appl Neurophysiol 50:159-167.

113. MORINGLANE JR, GRAF N, OSTERTAG CB (1987) Neurologia (Esp.) 2:202-210.

114. GODANO V, FRANK F, FABRIZI A, FRANK RICCI R (1987) Child's Nerv Syst 3:85-88.

115. BENABID A, BLOND S, CHAZAL J, COHADON F, DAUMAS-DUPORT C et coll (1985) Neurochir 31:295-301.

116. PECKER J, SCARABIN JM, BRUCHER JM, VALLEE B (1979) In: Laboratoires Pierre Fabre Publishers, Démarche Stéréotaxique en neurochirurgie tumorale, Paris.

117. LUNSFORD D, MARTINEZ J (1984) Surg Neurol 22:222-230

118. SCERRATI M, ROSSI GF (1984) Acta Neurochir (Wien), S33:201-205.

119. BENABID A, REMY C, CHAUVIN C (1986) In: Walker MD & Thomas DGT (eds), Biology of brain tumour, Martinus & Nijhoff, Boston, pp 221-236.

120. BACKLUND EO, GRANBERG PO, HAMBERGER B, KNUTSSON E, MATTENSSON A, SEDVALL G, SEIGER A, OLSON L (1985) J Neurosurg 62:169-173

122. BACKLUND EO, GREPE A, LUNSFORD D (1981) J Neurosurg 62:169-173.

121. BACKLUND EO (1987) Acta Neurochir 41:46-50

Cerebral gliomas. G. Broggi and M.A. Gerosa, editors

STEREOTACTIC INTERSTITIAL RADIOTHERAPY FOR GLIOMAS

CHRISTOPH B. OSTERTAG
Abteilung für Stereotaktische Neurochirurgie
Neurochirurgische Universitätsklinik
D - 6650 Homburg/Saar

INTRODUCTION

Since all treatment modalities in gliomas remain palliative and complete cure is realy achieved, the preservation of the intregrity of the brain is one of the essentials of every treatment. The integrity of the brain cannot be preserved by gross resections. The integrity of the brain is endangered by external beam radiotherapy depending on the total dose, the number of fractions, the irradiated volume and the age of the patient. Low dose rate interstitial radiotherapy offers a favourable alternative for tumors which are circumscribed, not resectable and slow growing.

Radiation from an interstitially implanted source such as Iodine-125 is continuously delivered at comparatively low dose rates compared with the dose rates delivered by external beam irradiation. Continuous low dose rate irradiation seems to increase the effectiveness of radiation possibly based on both, cell kinetic interactions and oxygen dependancy. The therapeutic ratio of interstitial radiotherapy is enhanced as a result of the rapid dose fall off in tissue within a distance of milimeters. Since radiation from Iodine-125 sources is effectively attenuated by by interjacent tissue such as brain, bone and scalp, radioprotection no longer poses a serious problem.

PATIENTS AND METHODS

From May 1984 until June 1988 151 patients habouring cerebral gliomas have been treated with Iodine-125 interstitial radio-therapy. All patients presented with progressive neurological deficits or had evidence of CT-controlled tumor growth or both. The majority were astrocytic tumors or mixed oligoastrocytomas (132 cases) (Tab.1). Anaplastic astrocytomas and ependymomas were less frequent. The majority of the lesions were located in the basal ganglia - thalamus, in the adjacent insula and temporal lobe and in the fronto- parietal lobes (Tab.2). Lesions in the hypothalamus were mostly hypothalamic gliomas or chiasmal gliomas with documented growth. 71 lesions were on the left, 51 lesions on the right hemisphere and 29 lesions in the midline of the brain. Twentynine patients were younger than 16 years of age. At the time of implantation 116 patients had a Karnofsky performance status of 70 or higher, 35 patients had a Karnofsky performance status lower than 70. The primary and presenting symptom was in 82 patients epilepsy, in 30 patients a paresis, 15 patients presented with increased intracranial pressure, 11 had ophthalmologic symptoms and 3 patients presented primarily with mental disturbances only. Diffusely infiltrative tumors crossing the midline, highly malignant, non delineated tumors such as lymphomas, glioblastomas, germinomas and multiple metastases were excluded on grounds of their invasiveness and/or tendency for seeding. Also excluded were patients with a rapidly deteriorating neurological status or major mass effect with eminent brain herniation.

After the implantation consecutive CT control scans were performed at 5 days, 6 weeks and every 3 months following the implantation. The patients were closely followed by neurological examination including Karnofsky performance status and compared with previous examinations.

DOSIMETRY

Iodine-125 seeds were used either as temporary (41 cases) of permanent implants (110 cases). The clinical dosimetry for permament Iodine-125 implants was carried out with respect of the physical and biological properties: Iodine-125 is emitting primarily a low energy X-ray radiation of 27 -35 KeV (1). According to several investigators, the relative biological efficiency seems to be in the range of 1.2 - 1.4. According to SONDHOUSE (11) the specific dose rate factor is 1.32 cGy/h and mCi at 1 cm in tissue. For a clinical situation the calculated accumulated dosage for permanent implants ranged between 2000 and 9000 cGy. This dose was calculated to accumulate within 90 days on the outer radius of a tumor when permanent implants were used. The mean dose rate varied between 2 and 10 cGy/h

(Tab. 3). Before the implantation of the sources each seed was calibrated in a ionization chamber (Curiementor,PTW, Freiburg). Because of the seed configuration, i.e. a titanium rod with welded ends, the dose distribution is unisotropic which makes the calibration difficult. The unisotropic dose distribution is, however, more or less irrelevant for the biological effect as demonstrated experimentally (3,9,10) and on autopsy material (5).

IMPLANTATION TECHNIQUE

Patients selected for interstitial radiotherapy have been undergone a previous stereotactic diagnostic procedure, i.e. computed tomography and cerebral angiography under stereotactic conditions and serial biopsies in one or more tracts establishing a morphological diagnosis including the delineation and invasiveness of a tumor,the composition (solid, cystic or necrotic), the vascularity and the classification (6,8). In patients where the diagnosis, delineation and classification was unequivocal, the implantation of the Iodine seed was carried out in one procedure. When there were doubts left regarding the diagnosis and grading a second procedure was required. For the diagnostic and interstitial radiotherapy procedure, a Riechert stereotactic system (Codman GmbH, Hamburg, West-Germany) was used. The patients were kept under local anaesthesia with a mild sedative medication. A modified CT table was used for CT scanning, angiography and operative procedure. The coordinates of the target volume are derived directly from the CT-image information (Somatom DR, Siemens Erlangen). Target coordinates for interstitial implants are usually choosen in the center of the tumor volume or along a suitable axis through the tumor. Exceptionally multiple targets and tracts were choosen. The number and the activities of the Iodine seed are selected according to the tumor shape and volume with special reference to the anatomical site. Whereas permanent seeds were used as lost implants, temporary seeds were kept in the tip of a heat shrinking teflon catheter which was secured subcutaneously in the burr hole. The catheters were removed by reopening the scalp incision after 30-60 days. Radiation protection regulations limit the level of radiation at discharge of the patient to less than 1 μSv per hour at 1

Tab. 1:
Interstitial Radiotherapy for Gliomas
(151 patients):

Diagnoses (WHO)	Cases
Pilocytic Astrocytoma (I)	28
Astrocytoma (II)	69
Oligo-Astrocytoma (II)	35
Anaplastic Astrocytoma (III)	12
Glioblastoma	4
Ependymoma (II)	3

4/84 - 6/88

Tab. 2:
Interstitial Radiotherapy for Gliomas
(151 patients):

Location	Cases
Frontal Lobe	38
Temporal Lobe - Insula	34
Parietal - Occipital Lobe	16
Basal Ganglia - Diencephalon	45
Ventricles	7
Brainstem	11

4/84 - 6/88

Tab. 3:
Interstitial Radiotherapy (Iodine-125) for Gliomas
- 151 cases

Calculated cumulative doses	5500 cGy (2000-9000)
Radius of irradiated volume	14.7 mm (5 - 25)
Implanted I-125 activities	9.9 mCi (.2 - 19.3)
Temporary versus permanent implants	41 : 110

4/84 - 6/88

meter distance. This level of radiation is rarely exceeded and special isolation of the patient in private rooms was not necessary.

RESULTS

The implants produced favourable responses resulting in stabilization in 80 patients or improvement of neurological symptoms in 45 patients for 3 to 60+ months which was paralled by CT-controlled reduction in tumor mass (Fig. 1). In close accordance with the neurological status the Karnofsky performance status was improved by more than 10 % in 49 patients. The status remained stable, i.e. a change of 10%, or unchanged in 80 patients. In 22 patients the Karnofsky performance status deteriorated by more than 10 %. In 21 patients we observed either an only short lasting improvement or no improvement at all. Two of these patients were recurrences after resection and external beam radiotherapy. These 21 patients neurologically progressively deteriorated, which required additional external beam radiotherapy in 11 patients. The patients who received additional external beam radiotherapy all had anaplastic astrocytomas. In 6 patients radionecrosis and continuous tumor growth made a craniotomy and resection necessary.

Twelve patients developed expanding tumor cysts 2-10 months after the implantation which required internal or external drainage over a Rickham-Reservoir. Since these cysts are part of the natural history of astrocytomas and were never observed in the immediate vicinity of the implant, it remains an unresolved question whether the cause for this are the interstitial implants or not.

Up to date 11 patients harboring anaplastic astrocytomas have died after 3-18 months (median survival 9.7 months). Because of the short observation period of this series data are preliminary in a patient group in which 86 % are still surviving. The calculated 5-year survival rate (life table method) was for astrocytomas WHO-grade II 78 % (standard error 7 %), for oligoastrocytomas 85 % (standard error 7 %), and for pilocytic astrocytomas 95 % (standard error 5 %). The 3-year survival rate for anaplastic astrocytomas was 35 % (standard error 20 %).

DISCUSSION

Growth characteristics, mitotic activity, vascularity and invasiveness of gliomas are major determinants for the effect of therapeutic intervention. In external beam radiotherapy the tolerance of the healthy brain is critical. It implies a therapeutic ratio i.e. a difference in radiosensitivity between tumor and brain. There is no doubt that external beam irradiation is effective in highly proliferating tumors such as glioblastomas. It's effectiveness is questioned for tumors with a very low proliferation capacity and no neovascularization. In those cases it seems wiser to remove as much tumor as possible. Most data generally support a philosophy of radical surgical removal of gliomas whenever possible. This conclusion, however, is biased that there is a tendency for the most favorable lesions to be treated with radical surgery and the less favorable, i.e. the deep-seated or infiltrative lesions to be treated with limited approaches, i.e. biopsy. When, however, there is no harmless way of surgical removal as in cases of gliomas in the basal ganglia, brain stem or in functionally critical cortical areas, stereotactic interstitial radiotherapy using low dose rate implants is considered an appropriate treatment modality. Biologically the local radionecrosis achieves a kinetically significant reduction of the proliferating pool of tumor cells. Although the technique of placing radioactive sources into the brain is not a new one, the inability to judge the effect of treatment and the tolerance of the surrounding healthy brain has hampered its development. Clinically tumor control was often incomplete and/or perifocal brain edema was a serious complication. In former years mostly high energy gamma- emitting sources like radium, cobalt-60, iridium-192 or gold-198 were used which have a more gradual radiation fall-off and thus effect tissue at great distance. Low energy gamma- emitting sources like iodine-125 allowed the delivery of high doses locally with a rapid fall-off over short distances. Attenuation predominates over scattering for distances greater than 1.5 cm. From experimental studies we have learned that iodine-125, iridium-192 and gold-198 implants can indeed effect sharply demarcated radionecroses in the brain. The implants also cause, like Yttrium-90, a disruption of the blood-brain-barrier with a consecutive vasogenic edema. The dose rate determined the volume of the necrosis and the magnitude of the vasogenic edema. The vasogenic edema determined the extent of perifocal demyelination and reactive gliosis. The amount of blood born fluids that cross the blood brain barrier depends on the total surface area of the irradiated defective capillaries and capillary permeability itself. To understand this phenomenon has far-reaching implications for clinical interstitial radiotherapy.

While a number of centers have used interstitial irradiation with high activity implants in conjunction with external beam irradiation and/or surgery for primary or recurrent malignant gliomas we have restricted its use to implants with low activity sources for non-malignant gliomas. Non-malignant differentiated gliomas, i.e. astrocytomas, oligoastrocytomas, oligodendrogliomas, ependymomas and moderately anaplastic astrocytomas pose problems for conventional radiation therapy. Current survival rates with external radiation therapy leave much to be desired. The data suggest that for incompletely resected lesions postoperative conventional radiotherapy improves 5 and 10 years survival rates by 20%. BLOOM correctly cautions against high dose and large volume brain irradiation "because of the possible risk of serious brain injury occuring before tumor recurrence" (2). Our results to date suggest that treatment of primary non-resected differentiated gliomas is feasible. The local tumor control by interstitial implants seems to be the least traumatic treatment modality although it, too, is considered palliative. The clinical data of the present series and data on survival while of interest, are preliminary. The larger Freiburg series, which now comprises more than 600 cases, quotes 5 years survival rates of 54 % for astrocytomas grade II, 43 % for astrocytomas grade III, 22 % for oligodendrogliomas grade II-III. These figures represent, however, very inhomogenous groups with respect to tumor location and size,age of the patient, clinical presentation and course (7). Combined interstitial and external radiotherapy was mainly used by the Paris Group. Both removable high activity iridium sources or local irradiation of the tumor volume and external radiotherapy for the greater tumor volume were used. That allowed the application of more moderate doses compared with external beam radiotherapy alone. Published 5-year survival data show a 78 % survival for gliomas grade I, 69 % for gliomas grade II and 55 % for gliomas grade III (12).

Our results are unsatisfactory in the treatment of anaplastic astrocytomas whether moderately or highly malignant. The biology of anaplastic gliomas is such as to make recurrence at the perimeter or elsewhere in the brain the cause of failure of this technique. Failure could even not be prevented with additional external beam therapy. GUTIN et al. (4) and SZIKLA et al.(12) also quote favourable results for supratentorial malignant gliomas, using combined interstitial and external radiation therapy. Although the importance of attempting local control of anaplastic astrocytomas or glioblastomas by surgery or radiation is strongly advocated, the author nevertheless disagrees with the tumor biologic concept of malignant gliomas as "localized" and therefore cannot recommend interstitial radiotherapy with Iodine-125-seeds for malignant gliomas.

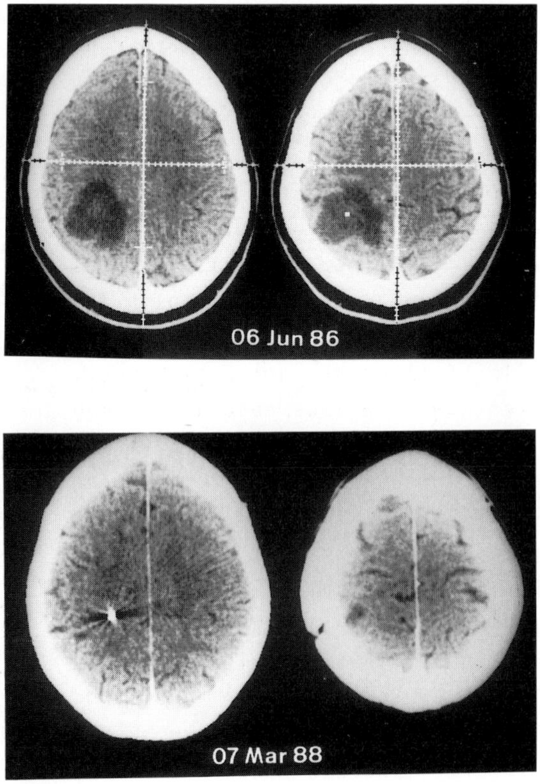

Fig. 1: Interstitial radiotherapy for a left parietal astrocytoma (WHO-grade II) in a 34-year old female (upper). Following a serial stereotaxic biopsy a permanent Iodine-125 seed was implanted into the center of the tumor (activity 15.15 mCi, total reference dose 6000 cGy on a diameter of 16.5 mm). CT-control (lower) revealed a shrinkage of the tumor without side effects or neurologic deficit (patient works as a secretary).

REFERENCES

1. Anderson LL, Hsin MK, Ing-Yuan D (1981) Clinical dosimetry with 125-I. In: Modern interstitial and intracavitary radiation cancer management. George FW (ed) Masson, New York

2. Bloom HJG, Walsh LS (1975) Tumors of the central nervous system. In: Bloom HJG, Lemerle J, Neidhardt MK (eds), Cancer in children. Clinical management. Springer, Berlin

3. Groothuis DR, Wright DC, Ostertag CB (1987) The effect of I-125 interstitial radiotherapy on blood-brain barrier function in normal canine brain. J Neurosurg 67:895-902

4. Gutin PH, Leibel SA, Wara WM, Choucair A, Levin VA, Philips TL, Silver P, Da Silva V, Edwards MSB, Davis RL, Weaver KA, Lamb S (1987) Recurrent malignant gliomas: survival following interstitial brachytherapy with high-activity Iodine-125 sources. J Neurosurg 67:864-873

5. Kiessling M, Kleihues P, Gassega E, Mundinger F, Ostertag ChB, Weigel K (1984) Morphology of intracranial tumors and adjacent brain structures following interstitial Iodine-125 radiotherapy. Acta Neurochir Suppl 33:281-289

6. Moringlane JR, Graf N, Ostertag CB (1987) Stereotaktische Diagnostik von Hirntumoren im Kindesalter als Grundlage für die Therapieplanung. Klin Pädiat 199:260-268

7. Mundinger F (1986) Stereotactic biopsy and technique of implantation (instillation) of radionuclids. In: Therapy of malignant brain tumors. Jellinger K (ed), Springer-Verlag Wien-New York

8. Ostertag ChB, Mennel HD, Kiessling M (1980) Stereotactic biopsy of brain tumors. Surg Neurol 14:275-283

9. Ostertag ChB, Weigel K, Warnke P, Lombeck G, Kleihues P (1983) Sequential morphological changes in the dog brain after interstitial Iodine-125 irradiation. Neurosurgery 13: 523-528

10. Ostertag ChB, Warnke P, Kleihues P, Bigner D (1984) Iodine-125 interstitial irradiation of virally induced dog brain tumors. Neurol Res 6:176-180

11. Sondhaus CA (1981) Physical properties, photon dosimetry and effectiveness. In: Modern interstitial and intracavitary radiation cancer management. George FW (ed) Masson, New York

12. Szikla G, Schlienger M, Blond S, Daumas-Duport C, Missir O, Miyahara S, Musolino A, Schaub C (1984) Interstitial and combined interstitial and external irradiation of supratentorial gliomas. Results in 61 cases treated 1973 - 1981. Acta Neurochir, Suppl 33:355-362

Cerebral gliomas. G. Broggi and M.A. Gerosa, editors

STEREOTACTIC INTERSTITIAL IRRADIATION OF SLOW GROWING BRAIN GLIOMAS: PRELIMINARY RESULTS

M. SCERRATI, R. ROSELLI, M. IACOANGELI, P. MONTEMAGGI (*), N. CELLINI (*)

Institute of Neurosurgery and (*) Institute of Radiology, Catholic University, Largo A. Gemelli 8, 00168 Rome (Italy).

INTRODUCTION

Stereotactic interstitial brachycurietherapy (BCT) represents a low dose rate irradiation able to produce selective radionecrosis of predetermined tumor volumes without affecting the surrounding healty tissue (1,2,3,4,5,6,7,8). Well differentiated brain gliomas present, as is well known, a thin therapeutic interval when treated with the conventional radiotherapy, i.e. their difference in radiosensitivity with the healty barin tissue is very small (3,4,6,7,8). This limit can be overcome by BCT which found in these tumors its best indication.

MATERIAL AND METHOD

Patients - From 1980 to 1987 25 patients (mean age 31 +/-11.3 years, 14 males and 11 females) with cerebral slow growing gliomas were treated with stereotactic brachycurietherapy. Histological diagnosis obtained with stereotactic biopsy was as follows: 1 pilocytic astrocytoma, 15 fibrillary astrocytomas, 6 oligodendrogliomas and 3 oligo-astrocytomas. The tumor volume to be treated, assessed by the integration of the neuroradiological data with the results of previous stereotactic biopsy (5,7,9,10), was less than 15cc in 6 cases, between 16 and 60cc in 14 and more than 60cc in 5.

Method - The stereotactic BCT was carried out as the sole treatment in 12 patients: it was combined with surgery in 4 cases (12 before BCT), with external radiotherapy in 7 (1 before BCT) and with both in 2. The radioactive sources utilized were Ir 192 in 21 patients and I 125 in 4. The sources were implanted permanently in 23 patients and temporary in 2. Tumor peripheral dose given by BCT in permanent implants ranged between 70-100 Gy in 17 cases and between 100-130 Gy in 6; as for temporary implants the given peripheral dose was 40-60 Gy in 5-7 days (8-12 Gy/24h).

RESULTS

The general follow up ranged between 0.3 and 6.9 years (m=2.5 yrs). The duration of symptoms before treatment ranged between 0.5 and 9.2 years (m=3.9 yrs).

Five years survival rate was 88% (+/- 6.5%). Three patients died within the second year since BCT owing to causes not related to the treated tumor (brain infarction contralateral to the implanted side, acute thyreotoxic syndrome and suicide respectively). All the other patients of this series are still alive, but 1 who died because of tumor progression 5.7 years after interstitial irradiation.

The Karnofsky performance status since the beginnig of BCT and independently from any precedeng treatmen, was evaluated only in the patients with at least 3 years follow up (11 cases). The initial score ranging between 0.60 and 0.70 rised to 0.80-1.00 in all patients during the first year after treatment and was lasting for many of them. A decrease of the scale rate followed in 4 patients who developed radionecrosis. The final score at 3 and 5 years follow up however never fell below 0.60, ranging in most of the treated cases between 0.70 and 0.90.

Epileptic seizures were present in all treated patients but one (24/25). Interstitial irradiation yielded a quick and stable decrease of seizure frequency in all patients, 2 of them being completely seizure free 5 years after treatment.

No mortality nor morbidity related to the stereotactic implant are affecting this series. Eight patients developed radionecrosis with a latency of 9-54 months (m= 19.7). In 4 of them a good clinical control was obtained with corticosteroids while in the remaining 4 surgery was necessary because of the extent of radionecrosis (2 cases) and owing to the tumor progression (2 cases) as documented with stereobiopsy.

DISCUSSION

A long term follow up is necessary to assess the therapeutic results in slow growing gliomas. Our series is too small and the period of observation too limited to allow conclusive considerations. Nevertheless some aspects can be remarked.

1 - Survival rate and quality of life seem to be satisfactory in particular when compared with the outcome of such types of tumor after surgery plus conventional radiotherapy, as reported in the majority of series (survival rate not superior to 30-40% of treated patients at 5 years) (6).

2 - The final goal of BCT able to give the best chance of cure, is tumor necrosis (1,3,7,8). To reach this goal a careful evaluation of the brain tolerance is required as to avoid the risk of healthy brain injury. Radionecrosis in fact, though often unprendictable, seems to some extent related to the dose/volume and dose/time ratio and to the biopathological aspects of the tumor.

3 - Permanent implants have been so far preferred in the treatment of slow growing gliomas, because of the very limited proliferating capacity of such tumors in the time. The majority of patients in our series (23/25) have been treated by this modality of irradiation. Recently we utilized temporary implants in 2 patients (Fig.1). The main advantages of temporary implants seem to be the control of the exact position of the sources, the better dose modulation especially when external radiotherapy has to be associated, and a more tolerable radionecrosis.

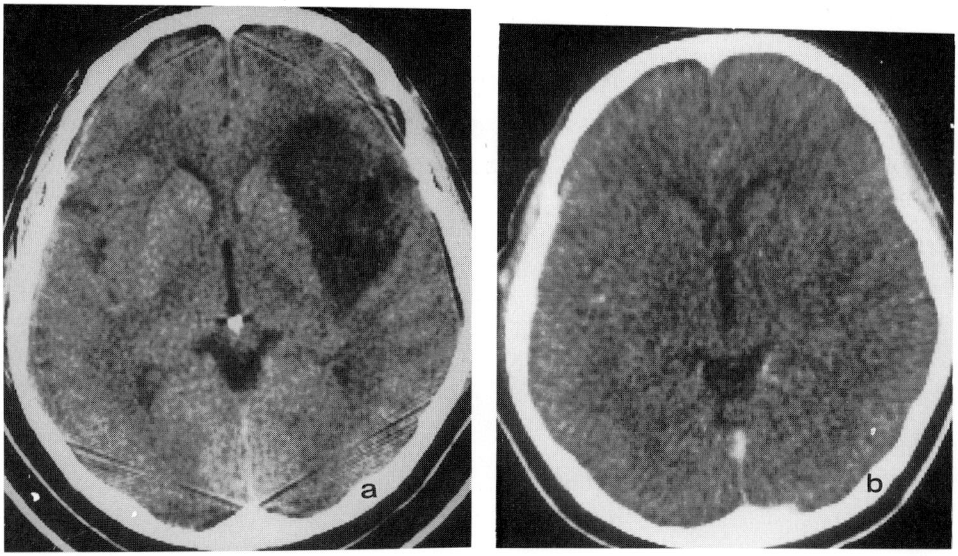

Fig.1. Right fronto-temporal oligodendroglioma before (a) and 18 months after BCT with temporary implant (b).

4 - Stereotactic brachytherapy of brain gliomas is to be considered a field still in progress and in continuous evolution. Choice of the sources and their modality of application, the definition of an optimal dose/time and dose/volume ratio represent nowadays still open problems to be faced in the coming future.

ACKNOWLEDGEMENTS

This research is partially supported by a grant of Italian Ministry of Public Education.

REFERENCES

1. Bernstein M, Gutin PhH (1981) Neurosurgery 9: 741-750
2. Gutin PhH, Bernstein M (1984) Progr Exp Tumor Res 28: 166-182
3. Mundinger F (1987) In: Jellinger K (ed) Therapy of malignat brain tumors, Springer, Wien-New York, pp 134-194
4. Mundinger F, Weigel K (1984) Acta Neurochir (Suppl 33): 367-371
5. Rossi GF, Scerrati M, Roselli R (1985) Appl Neurophysiol 48: 127-132
6. Sauer R (1987) In: Jellinger K (ed) Therapy of malignant brain tumors, Springer, Wien-New York, pp 195-276
7. Scerrati M, Arcovito G, D'Abramo G, Montemaggi P, Pastore G, Piermattei A, Romanini A, Rossi GF (1982) RAYS (Roma) 7:93-99
8. Szikla G, Schlienger M, Blond S, Daumas-Duport C, Missir O, Miyhara S, Musolino A, Schaub C (1984) Acta Neurochir (Wien) (Suppl 33): 355-362
9. Scerrati M, Rossi GF (1984) Acta Neurochir (Wien) (Suppl 33): 201-205
10. Scerrati M, Rossi GF, Roselli R (1987) Acta Neurochir (Wien) (Suppl 39): 28-33

© 1989 Elsevier Science Publishers B.V. (Biomedical Division)
Cerebral gliomas. G. Broggi and M.A. Gerosa, editors

LINEAR ACCELERATOR RADIOSURGERY OF CEREBRAL GLIOMAS

FEDERICO COLOMBO, MD

Department of Neurosurgery, City Hospital, Vicenza, Italy

SUMMARY

A clinical experience obtained by the employ of a radiosurgical technique utilizing a standard linear accelerator and a multiple non coplanar arc irradiation modality is presented. The series consists of 34 patients treated from November 1982 until March 1988. These patients harbored cerebral gliomas deemed not suitable to open surgery removal.

INTRODUCTION

Radiosurgery is the erogation in single session of high radiation doses from external sources into a target volume defined by stereotactic procedure (8). While inside the lesion a large overkilling dose is focused, nearby healty tissues are spared from radiation damage, owing to the steep dose gradient existing at the borders of the target.

Radiosurgical technique is employed to manage intracranial lesions unsuitable to open craniotomy procedures. It has found application for treating cerebral arteriovenous malformations and selected types of intracranial tumors (9,12).

In the original set-up, a dedicated apparatus consisting of 201 Co 60 sources arrayed on a spherical sector (Gamma Unit) is employed (1,8,9,12). Our technique utilizes a linear accelerator with a multiple, non coplanar, arc irradiation technique (2,4,5,6,7).

MATHERIALS AND METHODS

Criteria for eligibility to radiosurgical treatment were:
1) Location (deep or in highly functional areas)
2) Clear-cut borders at CT
3) 3D spherical shape
4) Dimensions less than 35-40 mm in diameter.

For the operative procedure, the patient is fixed to the head frame of our original stereotactic apparatus(3). Target localization is performed by CT in stereotactic conditions (7).

Multiple bioptic sampling is performed with the aim of nature, grading and tridimensional shape confirmation (4).

A head frame adaptor is fixed to the treatment couch of our Varian 4 MV linear accelerator. The center of target volume is made to coincide with the isocenter of the linear accelerator by controlled movements of treatment couch. Multiple arc irradiations are performed, always focused on the target, on different planes obtained by rotation of the couch around a vertical axis passing through the isocenter. High radiation doses (from 10 to 50 Gy) are delivered in single or double session.

In the follow-up period, patients were periodically assessed by clinical and tomografic evaluation at 1,2,3,6,9,12 months and then, each 6 months.

RESULTS

Since the introduction of the technique in clinical practice, 168 patients have been treated (4,5,6) (TABLE I). Among them, 34 patients harbored cerebral giomas deemed not suitable to open surgery removal.

TABLE I

LINEAR ACCELERATOR RADIOSURGERY

November 1982 - June 1988	
Arteriovenous malformation	83
Venous angioma	2
Cavernous angioma	1
Glial Tumors	
Low-grade astrocytoma	16
Anaplastic astrocytoma	15
Glioblastoma	1
Oligodendroglioma	2
Ependymoma	2
Choroid plexus papilloma	2
Pinealomas	
Germinoma	7
Pineocytoma	1
Pineoblastoma	1
Acoustic neuroma	8
Craniopharyngioma	8
Meningioma	8
Metastasis	7
Pituitary adenoma	1
Lymphoma	1
Medulloblastoma	1
MB Parkinson (VL thalamolisis)	1
	168

Low-grade gliomas

16 patients affected by low-grade gliomas were treated with

doses from 12 to 50 Gy. Lesions diameter varied from 12 to 30 mm. Mean follow-up is 23.4 months(from 58 to 5). 3 patients died for their disease (at 38, 21 and 14 months after treatment).

A clinically significant improvement has been obtained in 6 patients. In 5 patients clinical examination is stabilized. 2 patients displayed clinical deterioration due to their disease.

In 6 cases CT follow-up showed a marked reduction or a complete shrinkage of tumor bulk, often following a transitory mass increase(usually in months 4 to 9). In 5 cases no significant change was demonstrated at CT.

Anaplastic gliomas

15 patients harboring anaplastic gliomas were treated with single or double doses from 16 to 40 Gy. Lesions diameter ranged from 11 to 35 mm. Mean follow-up is 22.5 months (from 3 to 36).

8 patients died for their disease without showing any clinical or tomographic improvement from 6 to 16 months (mean 11 months) after treatment. 5 patients showed a progressive deterioration and were operated on (3 in our department), from 3 to 8 (mean 5.7) months after irradiation.

3 patients showed a significant and long lasting (24, 29 and 36 months) improvement. 2 of these patients displayed at CT follow-up a transient tumor increase from 6 to 12 months, followed by progressive mass shrinkage.

In 2 patients clinical and neuroradiological conditions are unchanged after more than 2 years.

Other glial tumors

1 glioblastoma was treated with 18 Gy. Clinical and ct picture remained unchanged for 15 months and after recurrence took place. 2 oligodendrogliomas were treated with 18 and 25 Gy without any detectable effect.

PATHOLOGICAL FINDINGS

One patient affected by a right thalamic grade I astrocytoma repeated bioptic study at the moment of maximum size increase 6 months after treatment. A typical fibrillary astrocytoma was changed to a largely necrotic lesion , surrounded by a wall in wich giant, often poynucleated, cells were prominent. Tumor vessels displayed endoluminal trombosis, endotelial proliferation and sometimes fibrosis of the vessels wall. Similar features,also if less prominent,were evident in operative specimens of operated patients.

DISCUSSION

Also if radiosurgery is able to provide a steep dose

fall-off at the edge of target volume - a prerequisite of stereotactic irradiation of cerebral tumors - nevertheless it has only episodically been employed in the treatment of cerebral gliomas (12). For these oncotypes, implantation of isotopes has been preferred and has gained a much wider acceptance (10). Since both techniques aim to the radioinduced "ablation" of tumor volume, it seems that there is no biological advantage in protracted irradiation afforded by isotopes implantation. Also pathological evidence obtained from our cases points to a biological evolution - in which vascular damage might play an important role - similar to that described in experimental works done in implanted cases (11).

Even if isotopes implant gives a steeper dose gradient allowing treatment of slightly larger tumor volumes (up to 40 mm in diameter), radiosurgery affords a more uniform distribution without any surgical trauma. With implanted sources the risk of seeds malposition or shift can not be avoided. Moreover, in protraced irradiation, a significant decrease of tumor volume can lead to inappropriate irradiation to healty tissue that become enclosed in the isotope range.

Irregular 3 D targets require more complicated therapy planning with multiple source implantation. On the other side,in radiosurgical technique, target volume can be adapted to irregular tumors by modifying the movement of X rays source, as some of our researches seem to suggest (5).

In conclusion, when the major problem of radiosurgical procedure (the choice of the dose) will be definitively solved, radiosurgery will become an appealing alternative to isotopes implantation in the treatment of cerebral gliomas.

REFERENCES

1. Arndt J., Backlund E.O., Larson B., Leksell L., Noren G., Rosander K., Rahn T., Sarby B., Steiner L., Wennerstrand J.(1979):Stereotactic irradiation of intracranial structures:physical and biological considerations. In **Stereotactic Cerebral Irradiations**. G.Szikla Ed., Elsevier, Amsterdam, pp 81-92

2. Chierego G., Marchetti C., Avanzo R.C., Pozza F., Colombo F.(1988): Dosimetric considerations on multiple arc stereotactic radiotherapy. **Radiotherapy and Oncology,** in press.

3. Colombo F., Zanardo A.(1984): Clinical employ of an original stereotactic apparatus. **Acta Neurochir.** supll. 33, 569-573

4. Colombo F, Casentini L, Benedetti A, Pozza F (1984): Biopsia e radioterapia stereotassica dei gliomi cerebrali. **Minerva Med.** 75:1327-1331

5. Colombo F., Benedetti A., Pozza F., Avanzo R.C., Marchetti C., Chierego G., Zanardo A. (1985): External stereotactic irradiation by linear accelerator. **Neurosurgery,** 16, 154-160

6. Colombo F., Benedetti A., Pozza F., Zanardo A., Avanzo R.C., Chierego G., Marchetti C. (1985): Stereotactic radiosurgery utilizing a linear accelerator. **Appl. Neurophysiol.,** 48, 133-145

7. Colombo F., Benedetti A., Pozza F., Avanzo R.C., Chierego G., Marchetti C.(1987): Linear accelerator radiosurgery in **Advanced technology in neurosurgery.** P. Pluchino, G. Broggi Eds., Springer. Heidelberg, pp 170-187

8. Leksell L.(1974): **Stereotaxis and radiosurgery: an operative system.** Thomas, Springfield, Il., pp 1-99

9. Leksell L. (1983): Stereotactic radiosurgery. **J.Neurol.Neurosurg.Psychiat.,** 46, 797-803

10. Mundinger F.(1987): Stereotactic biopsy and technique of implantation of radionuclids. In **Therapy of malignant brain tumors.** Kurt Jellinger Ed, Springer Verlag, Wien-New York, pp 134-194

11. Ostertag CB, Weighel K, Warnke P (1983): Sequential morphological changes in the dog brain after interstitial irradiation. **Neurosurgery** 13:523-528

12. Steiner L: Stereotactic radiosurgery with the cobalt 60 Gamma Unit in the surgical treatment of intracranial tumors and arteriovenous malformations in **Operative neurosurgical techniques**.H. Schmidek Ed, Grune & Stratton New York pp 515-529

© 1989 Elsevier Science Publishers B.V. (Biomedical Division)
Cerebral gliomas. G. Broggi and M.A. Gerosa, editors

RADIOSURGERY IN GLIOMAS (MIDDLE-LINE TUMORS)

OSVALDO O. BETTI, M.D.; ROBERTO ROSLER, M.D.
Institutos Médicos Antártida, Rivadavia 4990, (1424), Buenos Aires, Argentina

INTRODUCTION

Jean Talairach's methodology consist in a rigourously logical employ of techniques in such a way that each surgical gesture is preceded by another that gives accuracy and precision to the next one. After a correct anamnesia, the CT scan allows us to corroborate the semiology of the particular case, identifying the tumoral image. After that, the angiography will be the first step, because the knowledge of the vascular structures is capital for two reasons: 1) to recognize the vascular lamina that shows the cortical fissures indicating the cerebral convolutions, and 2) to avoid damaging the vessels with the different diagnostic and therapeutic instruments. The next step consist in obtaining ventricular position and size. Nowadays, ventricular aspects are obtained by means of MRI which enables to reduce the number of ventriculographies to very selected cases.

With all these elements of neuroradiology, it is possible to obtain the individual morphogram defined as a collection of vascular-ventricular-cerebral unmagnified and undistorted data, all superimposable, carried out in one session, integrated into the patient's own contour. Over this material it is possible to add biopsic tracks at each level where the sample was obtained. The citopathological information is also added to the previous schema (solid tumor, infiltration areas, etc.).

With all this information at hand, which conforms a "working diagnosis" according to Hobsley's definition, we are in good condition to propose (or not) a therapeutical indication based on a logical procedure (2). If a therapeutical indication is proposed, all the strategic details must be studied on the diagnostic morphograms themselves.

METHODS

Stereotactic biopsy

The stereotactic biopsy is the last step in this diagnostic procedure oriented by CT scan and MRI, and guided by three-dimensional stereotactic studies which reduces risk to minimum. Orthogonal approaches are most preferable because we can get to know all anatomic structures through the brain. In simple oblicuity the anatomical information is reduced to one plane, and in double oblicuity (polar approach) it is just ballistic. Brain biopsies must offer the most complete information: thay must allow a correct appreciation of the tumor (spatial type and nature). The pathologist is present at the operating room because we consider it a condition sine qua non for the control and orientation of the biopsies. In each track the contiguous serial samples must inform about the spatial organization of the lesion, its nature and solid and infiltrating aspects of the tumor.

We use the ojective malignancy criteria of Daumas-Duport based on four parameters (nuclear anomalies, mytosis, necrosis and endothelial proliferation)(4). We must stress the concept that a tumor is composed of a solid part, often considered as the only real tumor (T), and the infiltrating peripherical area (i) so common in brain tumors. A tumor is the addition of both (T + i) and treatment must be projected for both. Underestimation of the extent of infiltration might explain the majority of cases of recurrence of the tumor.

Radiosurgery treatment

Radiosurgery employs the crossing of thin radiation beams into a target point, generating a spheroidal volume depending on the diameter of the beam, the angles covered in sagittal and coronal planes and the distance between the portals of entry. This kind of hyperselective irradiation is based on: precision, small volume, protection of structures surrounding the lesion. Our methodology of radiosurgery with linear accelerator combines Talairach's system with the angular movement of accelerator gantry in one sense, and the angular displacement of the stereotactic frame in the other. The X-ray beams (10 MeV) are provided by a Clinac 18, Varian (Palo Alto). We uses angles of 140^0 in the coronal plane diadems and 120^0 in the sagittal plane. If we consider that each degree represents one portal of entry, we use as a mean more than 1500 of them. The number of coronal irradiation planes vary from 7 to 17 with a mean of 11. The target volume is placed in the isocenter of the linac; this volume depends on the diameter of the secondary collimator that we have added to the first one of the gantry. We have different types of secondary collimators (from 6 to 20 mm diameter) which allows us to use from 1 to 6 of them in each case in order to obtain complex volumes by optimazing the shape, adapted to real tumoral volume to be treated. This technique allows to obtain a very rapid decrease of radiation penumbra (1).

The irradiation strategy based on the volume and spatial shape of the lesion consist in defining the coronal irradiation planes (diadems), their angles (coronal and sagittal) and target characteristics (simple or compound). When these parameters are defined a computerized program especially developed is used in order to obtain the relative dose distribution which allows us to consider the radiosensitivity of tissues surrounding the target volume. We use the 70-75% isodose level just around the lesion volume. The addition of small doses from each portal of entry from different directions, all of them convergent to the center of the lesion, produce a high dose central volume. For this reason, convergent directions are not arbitrarily chosen but related to the type and size of the lesion and radiosensitivity of the other surrounding structures. The simulation obtained by a computerized system (patient model and beam model) produces after interacting on them different possible results that we can consider, accept or change. At present, we are using complex volumes and shapes more often in order to controur all the target volume, sometimes extremely irregular. The photon beam like a scalpel must treat the solid tumor (T), and in several cases we employed combined approaches for complete oncologic treatment (T + i) using external classical irradiation or endocavitary substances for cysts, or surgery before radiosurgery (integrated therapy).

MATERIAL

During the period 1982-1987, from 104 biopsed tumors located in middle-line, from thalamus to brain stem and from infundibulum to pineo-tectal regions, 22 cases of gliomas were treated with radiosurgery with high energy linac (12 males and 10 females). Their ages varied from 1 to 49 years (mean = 19 years). Twelve patients were younger than 16 years with a mean of 7.5 years old. The clinical symptomatology shown in 13 cases was HIC, 2 cases with associated HIC and sensorimotor deficit. Hypothalamic syndrome in 3 cases. Two cases presented only sensorimotor syndrome and 2 others were found incidentally. The delay between the first signs and radiosurgery lasted a minimum of 1 month and a maximum of 48 months (mean = 18 months). Six cases had been priorly submitted to classical neurosurgery, 2 underwent subtotal resection, the rest only biopsy. Ventricular shunt was performed in 8 cases.

In all cases, a serial stereotactic biopsy was performed. Histological diagnosis of gliomas was obtained in all cases. In the brain stem, from 32 cases biopsied only 2 astrocytomas grade I were found. In the anterior third ventricle, from 19 cases biopsied, 9 were gliomas (5 astrocytomas grade I, 3 astrocytomas grade II, and 1 was an oligodendrocytomas grade II). In the posterior third ventricle, from 21 cases biopsied, 6 were gliomas: 1 astrocytoma grade I, 1 grade II, and 3 grade III. One case was an oligodendrocytoma grade I. In the brain-stem-thalamic region, from 6 biopsied cases, 2 were gliomas: 1 astrocytoma grade I, 1 astrocytoma grade II, and 1 oligodendrocytoma grade II.

RESULTS

Astrocytomas grade I-II: 15 cases. Follow-up: 13 cases. Among them, 1 patient died 4 years after treatment; 4 have been alive for more than 3 years. The others were treated during the last 3 years.

Astrocytomas grade III: 3 cases; 2 patients died before 12 months; the other one is alive after 56 months.

Oligodendrocytomas: grade I, 1 patient alive after 48 months; grade II, 1 patient alive after 30 months.

Oligoastrocytomas: grade II, 1 patient dead before 12 months; grade III, 1 patient dead at 14 months.

If we consider all the gliomas grade I-II (18 cases), the follow-up of 16 of them shows 4 cases dead and 9 alive after 3 years. In gliomas grade III (4 cases), 3 of them died and 1 is still alive after 3 years.

From a total of 22 patients (follow-up in 20 cases): dead: 7 patients (5 of them before the year, 1 at 14 months, 1 at 48 months). Still alive: 13 cases (from 12-18 months, 2 cases; from 30-36 months, 5 cases; from 48-54 months, 3 cases; up to 56 months, 2 cases; up to 69 months, 1 case).

DISCUSSION

Tumors of the brain stem, hypothalamus, thalamus and pineal-quadrigeminal regions share the same problems and this discussion can be applied to the tumors of all these regions (3,4,5,6,7,8,9). In their study of 25 cases published in Cancer, Mantrovadi et al. (6) insist on the fact that tumors in these locations are seldom biopsied and the majority of the patients are treated without a histological confirmation. Tumors treated without previous nature and grading diagnosis are useless both for the sake of knowledge and for the improvement of results.

For these tumors, deeply situated and sometimes with slight initial symptoms (due in most cases to hydrocephalus), treatment is at first shunting and afterwards the main therapy is external irradiation. Patients still alive, treated by radiotherapy show -if they are children- extremely short stature and growth hormone (GH) deficiency. For this reason, in the Sick Children Hospital of Toronto the patients do not receive radiotherapy because the efficacy of this treatment remains controversial (3). The concept of whole brain irradiation for grade III-IV gliomas is based on the autopsy studies which confirm the presence of the tumor which is larger than the tumoral area determined by clinical investigation. "In only one report an apparent benefit of whole brain irradiation over the limited field therapy was demonstrated" (6). Mantrovadi et al. said: "additional documentation of benefit of whole cranial irradiation is awaited", and concluded: "Based on this autopsy study and on the information available in the literature, we recommended only a limited field irradiation for brain-stem gliomas" (6). For these reasons, the selection of our cases is based always on a serial biopsy and subsequent knowledge of the tumoral nature and grading, mandatory condition for any treatment. In cases in which the biopsy is negative or doubtful no treatment at all is performed, only CT scan and MRI controls.

If tumors are small, with little or no invasive component -spatial type 1- and grade from I to III, a protocol of radiosurgery treatment is chosen. In cases of small volume tumors, spatial type 2, radiosurgery and external radiotherapy are performed (only 2 cases treated).

One important fact is the possibility of using radiosurgery with linear accelerator instead of after-loading implantation avoiding the bloody approaching ways and having the capacity of modulating the irradiation strategy in order to adapt the "therapeutic volume" to the tumor volume. We ignored how radiosurgery really acts, but we can infere that the difference in results when compared with classical protracted radiation therapy is due to a "time effect" when a single dose is used (Cf. in AVMs).

The little experience that we are showing in this paper is not sufficient for definitive conclusion about the results of radiosurgery in gliomas. We must await series with statistical value and with a longer follow-up.

REFERENCES

1. BETTI O.O., DERECHINSKY V.E., Hyperselective encephalic irradiation with linear accelerator. Acta Neurochirurgica, Suppl. 33:385-390, 1984.

2. HOBSLEY M., The nature of clinical acumen. Theor. Surg., 1:10-18, 1986.

3. HOFFMAN H.J., STROINK A.R., DAVIDSON G., HENDRIK E.B., HUMPHREYS R.P., Pediatric Brain Stem Gliomas: Evaluation of biopsy. Concepts Pediat. Neurosurg., (Karger, Basel), 7:105-116, 1987.

4. KELLY P.J., DAUMAS-DUPORT C., KISPERT D.B., KALL B.A., SCHEITHAUER B.W., ILLIG J.J., Imaging-basal stereotaxic serial biopsies in untreated intracranial glial neoplasms. J. Neurosurg., 66:865-874, 1987.

5. LAWS E.R., TAYLOR W.F., CLIFTON M.B., ORAZAKI H., Neurosurgical management of low-grade astrocytomas of the cerebral hemispheres. J. Neurosurg., 61: 665-673, 1984.

6. MANTROVADI R.V.P., PHATAK R., BELLUR S., LIEBAER E.J., HAAS R., Brain-stem gliomas: an autopsy study of 25 cases. Cancer, 49: 1294-1296, 1982.

7. McLAURIN R.L., BRENEMAN J., ARON B., Hypothalamic gliomas: review of 18 cases. Concepts Pediat. Neurosurg., 7: 19-28, 1987.

8. PENDL G., Pineal and Midbrain Lesions. Springer-Verlag, Wien, 1985, 269pp.

9. PIEPMEIER I.M., Observation on the current treatment of low-grade astrocytic tumors of the cerebral hemispheres. J. Neurosurg., 67: 177-181, 1987.

10. SZIKLA G., BETTI O.O., BLOND S., Data on late reactions following stereotactic irradiation of gliomas. In: Stereotactic Cerebral Irradiation, INSERM, Symp. 12, Ed. G. Szikla, Elsevier North-Holland Biomedical Press, Amsterdam, 1979, 167-174.

Cerebral gliomas. G. Broggi and M.A. Gerosa, editors

IN VITRO PREDICTION OF CHEMOSENSITIVITY IN CEREBRAL GLIOMA

DAVID G.T. THOMAS and J.L. DARLING
Neuro-Oncology Section, Department of Neurological Surgery, The National Hospital, Queen Square, LONDON WC1. U.K.

INTRODUCTION

One of the chief hopes of experimental neuro-oncology has been that model systems employing cell culture, characterisation of cells in-vitro, and assessment of their biological behaviour in response to chemotherapy drugs, would provide more specific prognostic information about the outcome for individual patients. Classical neuro-pathology studies have shown that classification and grading of surgical specimens of malignant gliomas are important in the prognosis of brain tumour. However, the patient's age and the presence or absence of epilepsy, as a symptom in clinical series, are almost equally prognostic. It is likely that cell culture systems both in animal and in human brain tumour will show individual variation in biological response to chemotherapy and that these characteristics will be equally prognostic. In our laboratories studies have been carried out of chemosensitivity testing in animal and human brain tumours with in vitro and in vivo correlation.

CELL CULTURE METHODS

It has been known for over sixty years that human malignant gliomas are amongst the most easy to grow in tissue culture of all human solid tumours. This is probably due in part to an inherent aggressiveness of the tumour cells. It is also due to the absence of fibroblast contamination in the surgical biopsy specimen and the fact that bacterial common contamination, which is common for example in bowel tumours, is generally absent in surgical specimens from a neurosurgical operation. In a study by the authors and co-

workers (1) 83% of cultures grew successfully in-vitro. The resulting cells are astrocyte like and are aneuploid without evidence of fibroblast overgrowth (2, 3).
In-vitro chemosensitivity testing involves cell cultures, characterisation of the cells, drug treatment, and recovery of the cells after treatment with measurement of their residual viability (4). In order to make the laboratory methods practical for a clinical correlation several compromises have to be made. One has to vary the dose of chemotherapy agent used in-vitro, to overlap that likely to be found in-vivo. Further compromises have to be made to accommodate the probable cell kinetics and of brain tumour cells in culture and their metabolism of particular cytotoxic agents. Residual viability of cells in culture can be measured in several ways. The most generally used methods have been counting colonies resulting from cell division in the clonogenic assay (5), or to measure nucleic acid or protein synthesis in microtitration methods (3). The former method has the attraction that it is measuring directly proliferative and reproductive ability of tumour cells, however the methods are very time consuming and can be made effective in a relatively small proportion of individual cases. The latter methods may be automated, can generally be achieved in a relatively short period and are successful in a large fraction of the individual cases.

The authors and co-workers (1, 3) employ mechanical and enzymatic disaggregation of tumour biopsy samples to produce a single cell suspension which is used to establish the primary culture. Generally sufficient cells are grown within 14 days to enable subculture in microtitration of target tumour cells. These cells are exposed to graded concentrations of different chemotherapy agents. Arbitrarily drugs are added daily for 3 days followed by a recovery period of up to 14 days. Residual viability is estimated by auto-fluorography methods, which measure residual amino acid uptake using ^{35}S labelled methionine at periods up to 14 days after drug exposure. A value is determined which expresses the concentration of drug causing 50% inhibition of amino acid uptake compared with control (ID50). Large experimental populations, which may be termed a "training set", are used to specify a median ID50 for a particular class and grade of tumour. In this way the laboratory may report whether an individual tumour is apparently "sensitive" or "resistant" to a particular drug, when compared with the relevant similar tumour population at large.

ANIMAL STUDIES

A spontaneous astrocytoma in the mouse was described by Fraser (6) and subsequently modified for cell culture (7), and this has been used both in flank and intracerebral inoculation for correlation between clinical chemotherapy response in-vivo and in-vitro response. Several different lines with varying phenotypes have been derived. One line, designated 497P-1 is in-vitro relatively resistant to procarbazine but relatively sensitive to CCNU and vincristine. When this line is inoculated in the flank growth delay curves may be plotted in animals treated either with procarbazine, CCNU, or vincristine as single agents, or procarbazine, CCNU and vincristine as a triple agent regime. It is found that no growth delay is produced in-vivo by procarbazine where as the three other regimes produce significant growth delay. When the same tumour cell line is used for intracerebral inoculation and treatment given with procarbazine or the three alternative regimes it is found that procarbazine produces no significant increase in life span while the three alternative regimes do (8).

HUMAN STUDIES

The relationship between drug sensitivity in-vitro and clinical progress in patients with grade III and grade IV glioma has been investigated in a study of 117 patients treated at The National Hospital with a clinical protocol using the triple agent regime of procarbazine, CCNU and vincristine (PCV) following surgery and radiotherapy. The relapse free interval (RFI), which can be measured accurately, was chosen as the end point rather than survival because it is not affected by factors such as the use of high dose glucocorticoids, whether a patient is nursed at home or in hospital terminally, or the subsequent inclusion of patients into phase II trials at relapse. All correlations were carried out retrospectively and no attempt was made on the basis of the test to influence chemotherapy. Individual glioma cell cultures were assayed for sensitivity to each of the drugs used clinically using a ^{35}S-methionine uptake assay (1, 3). This assay, which involved the prolonged 3 day exposure to drugs followed by a 14 day recovery period is necessary to obtain stable ID50 values for most drugs, especially phase specific agents. In order to minimise internal variation in assay cultures were sampled after 3-5 days recovery, but before density limitation of growth occurred. The training set used to establish median values of ID50 for grade III and grade IV gliomas consisted of 200 biopsied cases tested in this way.

Statistical analysis was made of the correlation between apparent chemosensitivity in-vitro and clinical progress in the patients receiving the triple agent PCV treatment. Patients were designated either chemosensitive or non-chemosensitive to each drug using the cultured methods described above. Fourteen of 40 (35%) responded in-vitro to CCNU, 17 of 40 (42%) responded to procarbazine and 16 of 40 (40%) responded to vincristine. The length of relapse free interval was compared with those patients who were sensitive to a particular drug and with those who were not. Using the Mantel-Cox test, the differences in relapse free interval between those sensitive or not sensitive to procarbazine or CCNU in-vitro were significant (P=0.02 and 0.01 respectively). The differences between relapse free intervals in the case of sensitive or non-sensitive to vincristine were not significantly different. It appears that only sensitivity to procarbazine and CCNU in this test is associated with increased relapse free interval. Those patients who have undergone adjuvant chemotherapy were divided into 3 groups. Group 1 consisted of 22 cases sensitive in-vitro to either procarbazine and/or CCNU. Group 2 consisted of 18 patients who were apparently not sensitive to either of these drugs in-vitro. Group 3 consisted of 77 patients who had not been tested for chemosensitivity in-vitro. Group 1 remained relapse free for longer than the non-chemosensitive Group 2 (Mantel-Cox's test, P<0.0001). The relapse free intervals of those patients whose cells had been tested in-vitro (Groups 1 and 2) were compared with relapse free intervals for the untested Group 3 and no significant difference was found.

To examine the possibility that the improved relapse free interval in the apparently chemosensitive group could be related to other favourable prognostic factors a comparison was made between features of the groups. Sex, type or extent of operation, exact amount of radiotherapy treatment, degree of steroid cover and tumour site within the brain were not signficantly different in the 3 groups. However, there were differences in age and grade. Group 3, who were untested were older than the tested Groups 1 and 2. The patients who were apparently sensitive in-vitro to either CCNU or procarbazine (Group 1) tended to be younger than those patients who did not appear sensitive (Group 2). However, this difference did not reach statistical significance. There is no difference in the proportion of grade III and grade IV tumours between the tested Groups 1 and 2 and the untested Group 3. There were fewer grade III tumours in the non-sensitive Group 2 than in the sensitive Group 1. Thus, 16 of 22 (73%) apparently sensitive tumours were grade III but only 6 of 22 (27%) of sensitive tumours were grade IV. By contrast only 4 of 18 (22%) non-sensitive tumours were grade III, while 14 of 18 (78%) of these tumours were grade

IV. For Groups 1 and 2 Cox's proportional hazard model was fitted including in the model age, grade, site of tumour and chemosensitivity. The final analysis showed that when all other factors were taken into account, chemosensitivity was still prognostic of longer relapse free interval.

As a result of this pilot study a multicentre prospective randomised trial of adjuvant chemotherapy with the triple agent regime described above and incorporating in-vitro chemosensitivity testing has been started in the United Kingdom.

CONCLUSION

Both in animal and in human studies there is an indication that in-vitro prediction of chemosensitivity testing has relevance to the in-vivo behaviour of cerebral glioma. In the future it is hoped that in-vitro chemosensitivity testing will be of practical importance for predicting clinical response in individuals. It is not known why individual malignant gliomas, from an apparently homogeneous histological grade vary so widely in their apparent chemosensitivity in-vitro and in their response in-vivo. Basic scientific investigations will be required to determine at a cell biological level the differences between apparently similar tumours in their response to chemotherapy agents. A further possible use for in-vitro chemosensitivity prediction is in the screening and acquisition of new chemotherapy agents which may possess more specific and greater therapeutic effectiveness.

REFERENCES

1. Thomas, D.G.T., Darling, J.L., Paul, E.A., Mott, T.J., Godlee, J.N., Tobias, J.S., Capra, L.G., Collins, C.D., Mooney, C., Bozek, T., Finn, G.P., Arigbabu, S.O., Bullard, D.E., Shannon, N. and Freshney, R.I. (1985) Assay of anti-cancer drugs in tissue culture: Relationship of relapse free interval (RFI) and in vitro chemosensitivity in patients with malignant cerebral glioma. Br. J. Cancer. **51**:525.

2. Guner, M., Freshney, R.I., Morgan, D., Fresney, M.G., Thomas, D.G.T. and Graham, D.I. (1977) Effects of dexamethasone and betamethasone on in vitro cultures from human astrocytoma. Br. J. Cancer. 35:439.

3. Morgan, D., Freshney, R.I., Darling, J.L., Thomas, D.G.T. and Celik, F. (1983) Assay of anti-cancer drugs in tissue culture: Cell cultures of biopsies from human astrocytoma. Br. J. Cancer. **47**:205.

4. Darling, J.L. and Thomas, D.G.T. (1983) Results obtained using assays of intermediate duration and clinical correlations. In: Human Tumour Drug Sensitivity Testing In Vitro Techniques and Clinical Applications. (Eds. Dendy & HIll), London: Academic press, p.269.

5. Rosenblum, M.L., Gerosa, M.A., Wilson, C.B., Barger, G.R., Pertuiset, B.F., De Tribolet, N. and Dougherty, D.V. (1983) Stem cells studies of human malignant brain tumours: Part 1. Development of the stem cell assay and its potential. J. Neurosurg. **58**:170-176.

6. Fraser, H. (1971) Astrocytomas in an inbred mouse strain. J. Path. **103**:266-270.

7. Serano, R.D., Pegram, C.N. and Bigner, D.D. (1980) cell culture lines from a spontaneous VMDk murine astrocytoma (SMA). Acta Neuropath. **51**:53-64.

8. Bradford, R., Darling, J.L. and Thomas, D.G.T. (1986) The in-vitro and in-vivo chemosensitivity of VMDk murine astrocytoma cell lines. J. Neuro-Oncol. **4**:113.

Cerebral gliomas. G. Broggi and M.A. Gerosa, editors

STRATEGIES FOR ENHANCING DRUG UPTAKE IN GLIOMAS

William FEINDEL, Mirko DIKSIC, Lucas YAMAMOTO, Douglas ARNOLD, Eric SHOUBRIDGE, Jean-Guy VILLEMURE

McConnell Brain Imaging Center, Montreal Neurological Institute and Hospital, Montreal, Canada

ACCESS OF DRUGS INTO GLIOMAS

Over the past three years, we have concentrated on techniques to improve the tumor/brain differential uptake of drugs used in the treatment of malignant gliomas. Many factors play an important role in the access of drugs into gliomas, e.g. (i) Blood-brain barrier and blood tumor barrier (Intact in Grade I and II; Impaired in Grade III and IV gliomas), (ii) Peri-tumoral "edema" (capillaries compressed; low metabolism) and (iii) Tumor vessels (endothelial proliferation, thrombosis and necrosis, increased vascularity, arterio-venous shunt).

Most of these represent fixed structural or pathological changes. But among those that can be altered are the barriers between blood and brain, on the one hand, and blood and tumor, on the other hand.

We have examined the effect of intravenous hyperosmolar mannitol on the uptake in various intracranial tumors of gallium-68, a positron emitter that can be quantitated on serial PET scans (Fig.1).

The degree of change of the tracer uptake in the tumor as compared to normal brain was highest in glioblastoma multiforme and in one example of metastatic lung cancer. Gliomas of Grade I and II and one each of meningioma and epidermoid cyst showed no change. The advantage of the intravenous infusion of a hyperosmolar solution as compared to the intracarotid arterial

Classification	No. of Cases	Changes of Permeability by I.V. Mannitol	
		No. of Cases	Degree of Changes
Glioblastoma Multiforme	9	7	+ 50 to + 240%
		2	No change
Oligodendroglioma	1	1	+ 130%
Glioma Grade I and II	3	3	No change
Metastatic Cancer	1	1	+ 340%
Meningioma	1	1	No change
Epidermoid Cyst	1	1	No change

Fig.1: Changes in uptake of gallium-68 with infusion of mannitol.

infusion, centers on the maintenance of the blood-brain barrier in the normal brain while at the same time exploiting the disruption of the brain-tumor barrier (2,3).

Intracarotid arterial infusion of a hyperosmolar solution opens the blood-brain barrier in normal brain, thus reducing the differential effect already present between tumor and brain. The access and toxic effect of chemotherapeutic drugs to normal brain would then be increased.

ARTERIAL INFUSION INCREASES TUMOR/BRAIN RATIO OF DRUG

Drug effectiveness is also influenced by many different factors which include (i) Arterial vs. venous route, (ii) Hyperosmolar agents, (iii) Physical properties of drug, (iv) Cell cycle effect (specific: Methotrexate, non-specific: BCNU, (v) Toxicity of drug (to tumor, to other tissues), (vi) chemosensitivity of tumor cells, (vii) penetrability/mass effect, and (viii) Steroids.

Another means of improving the concentration of drug in a glioma depends upon the slow infusion of the chemical by means of a catheter in the carotid artery above the ophthalmic branch.

Failure to attain this critical placement of the catheter can result in retinal damage. A comparison of intravenous and superselective intra-arterial carotid infusion of labeled BCNU tracer, showed that the latter achieved concentrations of the drug in the tumor that averaged 50 times higher (4,5). In the same study the radioactivity cleared more slowly from tumor than from normal brain. This suggests that decomposition of the BCNU occurred at a higher rate in the tumor resulting in "metabolic trapping" of the labeled BCNU breakdown products. For a given dosage of BCNU, it would appear that superselective arterial administration results in greatly increased delivery of drug to the tumor and consequent sparing of normal brain from the chemical agent (6). The application of these findings are now being explored in regard to increased survival. Evidently, an important aspect in the selection of the patients is to define tumors that lie within the vascular territory of the perfused carotid artery. This can be checked by injecting a labeled tracer dose of the drug and following with a PET scan to correlate the uptake of the drug within the tumor volume as displayed on CT or MRI (5).

PET STUDIES OF TUMORS

Positron emission tomography (PET) and labeled ^{11}C-BCNU, used in a series of studies on patients with malignant gliomas, gave the following results.

There is a dissociation between the metabolic rate for oxygen and that for glucose. In some gliomas the glucose utilization was increased as compared to that of the contralateral normal cerebral cortex. At the same time there could be 80% reduction of the cerebral metabolic rate for oxygen indicating abnormal glycolysis.

Cerebral blood flow within the gliomas was generally reduced from 20% to 70% as compared to normal cerebral cortex. However, local hyperemia associated with increased cerebral blood volume was occasionally observed on the margins of the tumor.

The pharmacokinetic study of ^{11}C-BCNU indicates that the initial uptake of the radioactivity in the tumor area is proportional to cerebral blood flow. Within 20 minutes after intravenous injection of the labeled drug, the radioactivity in the tumor is already higher than that in normal cerebral cortex.

Clearance curves of the labeled drug were analyzed by serial PET studies and compared to chemical analysis of the drug breakdown products in the blood. This showed that while the clearance of the radioactivity from the glioma is slower than from normal brain, this does not seem to be dependent upon blood flow through the tumor. These results suggest that both chemical and metabolic BCNU "decomposition" are different in the tumor than in normal brain. These results were confirmed in studies on rats with implantation into the brain of AA ascites tumor. It was found that after injection with labeled BCNU, chemical analysis of the tissues demonstrated that the radioactive levels were 1.5 times higher in the tumor than in contralateral brain. Most of this was due to 2-chloroethylisocyanate, one of the cytotoxic breakdown products of BCNU, which binds to amino groups of nucleic acids.

Earlier studies using PET to measure the glucose activity in patients with malignant gliomas suggested that hypermetabolism of glucose correlates strongly with the tumor grade and survival, although some studies failed to show this correlation. Most of this earlier work was carried out on a mixed series of patients, some with newly appearing tumors and some with recurrent lesions previously treated by surgery, radiation, and antitumor drugs.

In gliomas which had not undergone treatment but were later confirmed histologically, there was a range of metabolic findings. The cerebral blood flow was variable, the cerebral blood volume usually high, and the cerebral metabolic rate of oxygen low. The rate of glucose metabolism in these pretreated gliomas was often lower than that of the contralateral gray matter. However, there was no increase in oxygen extraction to indicate that circulatory perfusion was inadequate to meet metabolic needs. These findings indicate an intrinsic metabolic difference between pretreated gliomas and those recurring after therapy in which relatively high metabolic values have been reported (7).

A further useful application of the PET system for investigating brain tumors is the advantage of quantitating the pharmacokinetics of a labeled drug to determine its concentration and persistence within the tumor as compared to the normal brain. The baseline studies carried out using labeled BCNU were compared in patients with those defined by PET scanning for SarCNU, another chloroethylnitrosourea with a sarcinamide side arm. This particular drug was examined because it was more effective in reducing the cloning growth derived from human gliomas in tissue culture and was 10 times less myelotoxic in vitro, as compared to BCNU.

Kinetic studies with PET carried out on successive days in the same patient with a malignant glioma showed an important increase in the differential between uptake of SarCNU in the tumor as compared to normal brain.

PHOSPHORUS-31 MAGNETIC RESONANCE SPECTROSCOPY (PMRS)

We have used phosphorus magnetic resonance spectroscopy (PMRS) to monitor the metabolic changes of gliomas in vivo after chemotherapy with BCNU. Important changes may occur within hours of drug administration, long before structural changes appear on CT or MRI (8)

An example is provided by a 40 year old patient who had a recurrent mixed glioma which was judged to be Grade II histologically. As a means of monitoring infusion of BCNU delivered by way of a carotid catheter into the middle cerebral arteries supplying the tumor, the patient first had an imaging study and then spectroscopy, performed on a 1.5 tesla Philips gyroscan. The spectrum was derived from a block shaped volume of interest that included the tumor mass (Fig.2).

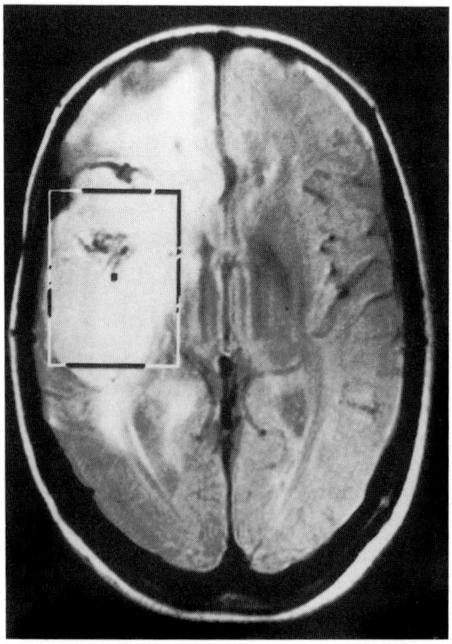

Fig.2: MRI scan with volume of interest shown by rectangle (TE=50 msec., TR=1200 msec.)

Before drug administration the phosphodiesters were 25% less than in normal brain. The intracellular pH was 7.14 (Fig.3). (The normal intracellular pH of human brain is 6.97 ± 0.02.) Eight hours following treatment, phosphocreatine and phosphodiesters were reduced by 40% while the intracellular pH was up to 7.24 (Fig.4). Thirty-two hours after treatment there was a recovery in the peaks of the phosphocreatine and phosphodiesters to within 20% of the control, but the pH continued to increase to 7.35 (Fig.5).

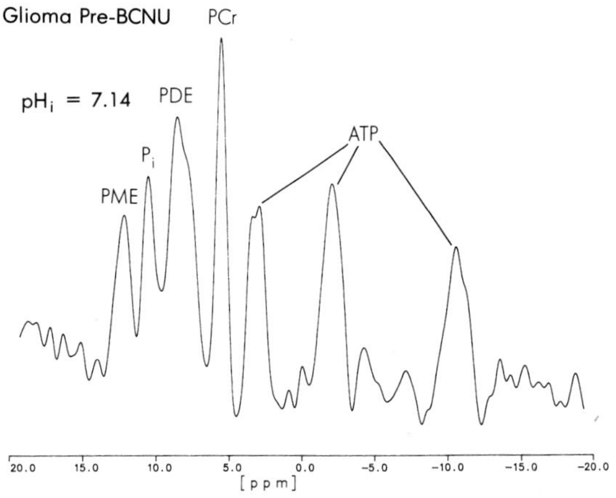

Fig.3: PMRS showing high intracellular pH and reduced PDE

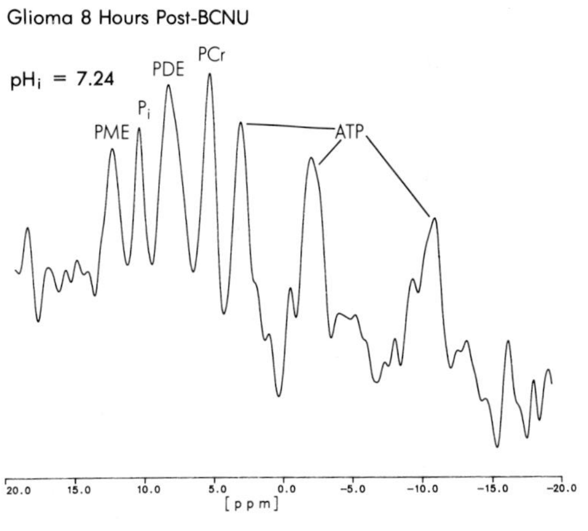

Fig.4: PMRS after intracarotid BCNU showing reduced PCr and PDE and higher pH

Fig.5: PMRS showing recovery of PCr and PDE peaks, but higher pH.

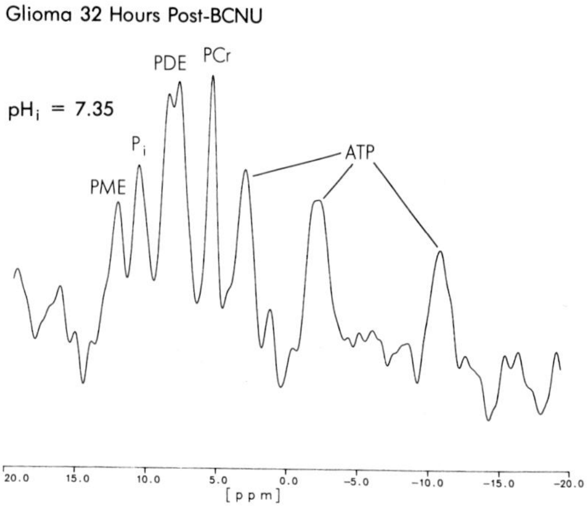

This illustrates the capability of PMRS for monitoring by a non-invasive repeatable method, the metabolic and pH changes in the tumor in response to intra-arterial antitumor drugs. The finding of an alkaline pH which becomes more pronounced after the intra-arterial drug infusion is of particular significance. The drug BCNU breaks down both in vitro and in vivo more rapidly with increasing alkalinity (9). Thus the alkaline range of pH in the tumor would favor this rapid breakdown and consequently more entry of the drug into the tumor. Our preliminary studies have shown that this increasingly alkaline pH does not follow intravenous administration of the same drug, thus suggesting a differential effect on the intra-arterial delivery of the BCNU (10). It is possible that the pretreatment pH may prove a useful predictor of chemotherapeutic response or toxicity to the tumor. The fact that acute metabolic changes occur within hours of the intra-arterial infusion of the drug suggest that magnetic resonance spectroscopy has promise as a means of early monitoring of therapeutic efficacy.

CONCLUSIONS

1. PET can be used as a method to study therapeutic techniques for gliomas. Using labeled cytotoxic drugs such as BCNU, it is possible to analyze and quantitate the speed of entry, concentration and persistence of the drug in the tumor and normal brain.

2. Uptake of a drug in the higher grade gliomas is increased after hyperosmolar intravenous infusion.

3. Intra-arterial carotid supraophthalmic infusion produces a higher concentration of drug than intravenous administration.

4. Identity of the vascular territory within which the glioma is located can be ascertained by carotid infusion of labeled tracer amount of the drug.

5. Metabolic activity and blood flow of gliomas before and after treatment can be examined by PET. Studies on pretreated gliomas show no consistent metabolic parameter that can indicate grade of malignancy of the tumor. The metabolic profiles of gliomas, whether treated or pretreated, appear to be too variable to be useful as indicators of tumor response to treatment.

6. The pharmacokinetics of antitumor drugs can be examined in their labeled form by using PET, as, for example, comparison of BCNU and SarCNU. (11)

7. Phosphorus-31 magnetic resonance spectroscopy offers the advantages of combining detailed structural MR imaging with the spectrum of phosphate metabolites. Early changes in the spectral pattern indicate a metabolic response of the tumor to chemotherapy.

8. Using PMRS, malignant gliomas have been found to have an alkaline pH as contrasted to a lower pH for Grade I and II gliomas.

9. Intra-arterial but not intravenous BCNU increases the alkalinity of the pH in the tumor. This alkalinity has relevance to the rapidity of the breakdown of the parent BCNU molecule into cytotoxic components.

ACKNOWLEDGEMENTS

This work was supported by Grant No.NS 22230-03 from the Department of Health and Human Services, United States Public Health Services and by the Killam Research Fund and the Clive Baxter Research Fund of the Montreal Neurological Institute.

REFERENCES

1. Villemure JG, Yamamoto LY, Diksic M, Feindel W (1982) Annual Report, Montreal Neurological Institute and Hospital.

2. Yamamoto YL, Diksic M (1985) In: Reivich M (ed.) Positron Emission Tomography. Alan R. Liss, Inc., pp.413-423.

3. Rapaport SI, Hori M, Klatzo I (1972) Am J Physiol 223: 323-331.

4. Diksic M, Sako K, Feindel W, Kato A, Yamamoto YL, Farrokhzad S and Thompson C (1984) Cancer Research 44: 3120-3124.

5. Tyler JL, Yamamoto YL, Diksic M, Théron J, Villemure JG, Worthington C, Evans AC, Feindel W (1986) J Nucl Med 27: 775-780.

6. Fenstermacher JD, Cowles AL (1979) Cancer Treat Rep 61: 519-526.

7. Tyler JL, Diksic M, Villemure JG, Evans AC, Meyer E, Yamamoto YL, Feindel W (1987) J Nucl Med 28: 1123-1133.

8. Arnold DL, Shoubridge EA, Feindel W, Villemure JG (1987) Can J Neurol Sci 14: 570-575.

9. Montgomery JA, James K, McCaleb GS et al (1967) J Med Chem 10: 668-674.

10. Arnold DL, Shoubridge EA, Villemure JG, Feindel W (1988) 7th Ann Meeting, Soc Mag Res in Med. August

11. Skalski V, Rivas J, Panasci L, McQuillan A, Feindel W (1988) Cancer Chemother Pharmacol. (In press)

© 1989 Elsevier Science Publishers B.V. (Biomedical Division)
Cerebral gliomas. G. Broggi and M.A. Gerosa, editors

INTRATUMOR DRUG PERFUSION

RICHARD D. PENN
Department of Neurosurgery, Rush-Presbyterian-St. Luke's Medical Center, 1653 W. Congress Parkway, Chicago, Illinois 60612

INTRODUCTION

Traditional chemotherapy of brain tumors has failed. The reason may be that the agents which have been tried are not active against gliomas, or it may be that they have not been delivered in adequate amounts for long enough times. Direct infusing of drugs into brain tumors offers the possibility of bypassing the blood-brain barrier and testing the real activity of drugs. To understand the potential of this technique, it is necessary to review briefly the problems with intravascular delivery, the way in which chemicals move within the brain substance, and the experience to date with direct infusion in animal and human experiments.

PROBLEMS OF THE VASCULAR ROUTE

The key problems of vascular delivery of drugs are uneven distribution and poor penetration of water-soluble agents, poor effectiveness of the lipid-soluble agents so far tested, and limitations on the maximum concentration and exposure time that can be achieved. A further problem is that the intravascular route exposes the entire body to drug, and systemic toxicity becomes a major dose-limiting factor.

If a water-soluble agent is used intravenously, penetration into the normal brain parenchyma is prevented by tight junctions between endothelial cells. In high grade tumors, these junctions are often absent in their neovasculature, so leaks occur. However, the leakage is variable and drug distribution will not be uniform. Furthermore, in regions of necrosis no blood vessels may be present, so drug delivery is impossible. As a consequence, most clinical trials have used agents which are lipid-soluble and readily penetrate the blood-brain barrier. Unfortunately, the best drug to date, BCNU, has added very little to average survival times.

Other intravascular approaches are being tried. Modification of the blood-brain barrier by osmotic agents is one, but it has not yet been proven effective. The difficulty may be that osmotic disruption works as effectively on normal brain vessels as on tumor vessels. With barrier disruptions, the increase in drug delivery will be greater for normal tissue than a tumor because tumors already are partly "leaky". This will produce toxicity to normal brain tissue. Another alternative intravascular delivery method is direct arterial perfusion. The concentration of drug presented to the brain can be markedly increased by arterial intracranial catheterization. Also, exposure time can be lengthened by using external or internal implanted pumps. Thus, concentration and duration can be maximized for higher tumor cell killing. Obviously, only certain tumors with appropriate arterial supplies can be treated in this fashion, and further testing will be needed to demonstrate the advantages of this technique in carefully selected tumors.

ADVANTAGES AND DISADVANTAGES OF DIRECT PERFUSION

Direct perfusion seems to avoid many of the difficulties associated with intravascular administration. The ideal drug for intratumor infusion is water-soluble, so many of the agents used for solid tumors, like cis-platinum, can be given. Furthermore, high concentrations of drug can be maintained for long periods if drug pumps are utilized. Finally, relatively little drug has to be delivered into the tumor to produce high interstitial levels, so systemic toxicity is not a problem. As will be discussed, these advantages have to be balanced with distribution difficulties.

To understand how intratumor infusions can be used, the way in which substances diffuse through the brain must be considered. This is a very complex subject, but some general points can be made. The primary pathway substances move on through the brain is via the extracellular space. This space constitutes 20-30% of the brain volume, but is an irregular maze of small channels. Diffusion, rather than bulk flow of CSF, is the major force distributing chemicals in these channels. This is in contradistinction to the large CSF spaces which rapidly move substances by bulk flow propelled by the generation of CSF and vascular pulsations. Thus, a drug put into the ventricle rapidly goes throughout the large CSF spaces, but is very slow to penetration in the brain parenchyma.

Diffusion is a very ineffective means of transport. Just how poor has been demonstrated in a series of experiments by Fenstermacher and his colleagues (1,2). Stable, small molecular weight markers were introduced into the ventricles and the distance into the brain they moved was determined. Even with long perfusion times, the distances were only millimeters. For example, when methotrexate, a chemotherapeutic agent used to treat carcinomatosis of the meninges, was infused for four hours the concentration gradient from the ventricle to the brain tissue fell by ten-fold every two millimeters. This lack of penetration explains why intrinsic brain tumors cannot be adequately treated by ventricular perfusion. This is true even if the tumor is on the ventricular surface, since diffusion within a tumor is no better, and, in fact, may be worse than in brain tissue.

Another important point was made by Fenstermacher's group. Lipid-soluble drugs which are rapidly removed from the brain penetrate even less. Their measurements showed that BCNU decreases by a factor of 1000 just 2 millimeters from the ventricular surface. Thus, lipid-soluble drugs are the worst condidates for intraparenchymal use.

CISPLATIN EXPERIMENTS

The practical problems of intratumor perfusion are illustrated by our own attempts to use this technique over the last several years. The first step was to identify an agent which might be effective in treating gliomas. One drug which has shown considerable promise for solid tumors is cisplatin. Because it is water-soluble, it has not been used extensively for brain tumors. However, that characteristic plus its stability and small size suggested it would be good for direct perfusion.

In order to use this drug properly, the distribution of cisplatin when directly perfused into brain tissue had to be determined. Rats were implanted with a small catheter stereotaxically, and an Alzet minipump was filled with drug. The pumps delivered 0.9 μg/hr for seven days, after which the brain tissue was analyzed to see how far the cisplatin diffused. Experiments on cerebellar and cerebral tissue gave the same results. The concentrations fell rapidly from the center, where the catheter was placed to the periphery. By 5 millimeters, the concentration had decreased by a factor of 500. Using estimates of the concentrations necessary to treat glioma cells, we calculated that a single source of cisplatin could treat a 1 centimeter in diameter volume (3).

Next, we ran a series of experiments treating a well-characterized glioma model (4). With chronic intratumor perfusion, the 9L tumor growth was significantly slowed compared to controls. 5-FU, another water-soluble agent, was even more effective in this model. Neither agent showed any activity against the tumor when given systemically. Furthermore, the dose infused was much lower than with vascular treatment, so there were no systemic toxicities.

The problem in applying these findings to man is that the tumor volumes to be treated are much larger than in the rat. The simplest solution would be to use multiple catheters with many holes to act as point sources. This would be analagous to implantation of multiple radioactive beads. Unfortunately, this is a complex undertaking because many pumps have to be used. If two or more catheters are attached to a single pump, then the flow of fluid in each will depend upon resistance. The lower the resistance, the more the flow. To assure an even distribution, separate pumps have to be attached to each catheter. Since we had already determined that the average volume treated per catheter is only a centimeter, for most tumors many pumps are required. For a tumor 2x2x2 centimeters, 8 pumps are necessary, but for one that is 4x4x4, 64 pumps are needed.

As a feasibility study, we joined forces with a stereotaxic surgeon and an engineering group in Montreal, Canada and designed a multiple catheter system suitable for stereotaxic implantation (5). Alzet minipumps were attached to each catheter. They were preloaded with cisplatin and could pump for 10 days. The region of the tumor was determined from CT scanning, and arteriograms were done to avoid major vessels when the implant was performed. Several patients with recurrent glioblastomas were treated. After intratumoral infusion, all of the patients had brief remissions of up to six months and then deteriorated and died. A large number of pumps and catheters had to be used, 60-80 each time, to cover the large tumor volumes being treated.

What have we learned from this experimentation? First, the

logistical problems of intratumoral infusion can be overcome, but only with considerable difficulty. Multiple catheters and pumps are awkward and cumbersome, but they can be placed without injuring the patient. Second, the response to treatment is not long-lasting, so either better drugs have to be used, or the treatments must be repeated. Third, systemic toxicity can be avoided by intratumoral treatments, and are not dose-limiting as in the vascular route.

The future development of these techniques has to be both technological and biological. The multiple pumps could be avoided if, instead of placing catheters, a slow biodegradable polymer with a suitable chemotherapeutic agent were implanted. Lipid-soluble agents like BCNU are poor candidates because of lack of diffusion, but cisplatin or 5-FU should be tried. More importantly, biologically active agents which encourage cellular differentiation or enhance immunological responses should be considered. The important point to remember in application of these techniques is the very limited range through which chemicals may diffuse. For the rational employment of any agent for intratumor perfusion, careful laboratory studies will have to be performed to demonstrate the amount of volume that can be treated by that particular drug. Unless this information is known, successful application cannot be expected. However, the direct intratumoral perfusion opens the door to many more new agents, some of which may be of use in treating gliomas.

REFERENCES

1. Fenstermacher J, Kaye T (1988) In: Penn RD (ed) Annals of the New York Academy of Sciences: Neurological Applications of Implanted Drug Pumps. The New York Academy of Sciences, New York, vol 531, pp 29-39

2. Blasberg RG, Patlak C, Fenstermacher JD (1975) J Pharmacol Exp Ther 195:73-83

3. Kroin JS, Penn RD (1982) Neurosurgery 10(3):349-354

4. Penn RD, Kroin JS, Harris JE, Chiu KM, Braun DP (1983) Applied Neurophysiology 46:240-244

5. Bouvier G, Penn RD, Kroin JS, Beique R, Guerard MJ (1987) Neurosurgery 20(2):286-291

AUTHOR INDEX

SUBJECT INDEX